Excavation Safety

A Guide to OSHA Compliance and Injury Prevention

Carl O. Morgan

ABS Consulting
Government Institutes
Rockville, Maryland

ABS Consulting

Government Institutes
4 Research Place, Rockville, Maryland 20850, USA
Phone: (301) 921-2300
Fax: (301) 921-0373
Email: giinfo@govinst.com
Internet: http://www.govinst.com

06 05 04 03 4 3 2 1

ISBN 0-86587-959-1

Printed in the United States of America.

To the memory of

Russell Oscar and Kate Lucille Morgan

Contents

Chapter 7
Guide to OSHA Compliance and Hazard Control 55

List of Figures and Tables

Preface

This book addresses the most common causes of excavation, trenching, and shoring violations and provides readers with the insight they need to ensure compliance and protect workers lives and company investments. It provides step-by-step methods for restricting workers from hazardous areas, providing egress from excavations, protecting workers from materials and equipment, selecting competent people to perform self-inspections, and ensuring a protection system is in place.

Chapter 1 covers the initial planning and preparation considerations required to ensure adequate safety during excavations. The chapter provides an overview of planning for factors such as underground installations, trench design, soil conditions, and management's role in ensuring safety.

Chapter 2 explains soil characteristics and the reasons for—and dangers of—cave-ins. Type A, B, and C soil classifications are described, as well as methods to determine soil classifications. This chapter also covers signs of dangerous soil conditions and cave-ins.

Chapter 3 outlines the support systems that protect against cave-ins. Sloping, trench boxes, barricades, and other protective measures are explained. This chapter also provides information on protection from other hazards, such as falls, falling equipment, and hazardous atmospheres.

Chapter 4 covers the Occupational Safety and Health Administration (OSHA) requirement for safety training, including voluntary training guidelines, the training process, and proper identification of training needs. Determining when, where, and how much training is needed is explained in detail.

Chapter 5 provides information on self-inspections. Some of the topics covered include inspections by competent persons, inspections of protective systems and equipment, and the shared responsibilities of contractors and subcontractors.

Chapter 6 covers utility hazards. This chapter consists of a number of case studies of accidents involving underground utilities. Following a description of the accident, each section includes corrective actions, controls, and recommendations for making future excavations safe.

Chapter 7 is a guide to OSHA compliance and hazard control. This chapter provides the text of the three excavation standards of 29 Code of Federal Regulations (CFR) 1926 with important notes and tips for compliance added by the author. Each note is set off from the text in a box for the reader's convenience.

The appendices include an excavation condition log (Appendix A), typical site assessment questions (Appendix B), additional web resources (Appendix C), review questions on excavation safety (Appendix D), and a glossary.

About the Author

Carl O. Morgan is a Certified Safety Professional (CSP) with over 20 years of experience in general industry and construction safety compliance. Mr. Morgan is currently a safety engineer at the Los Alamos National Laboratory in Los Alamos, New Mexico. He is the author of Government Institutes book Crane Safety.

Acknowledgement

The author obtained much of the material for this book from OSHA's extensive website. For more information, please visit OSHA's website at www.osha.gov.

Introduction

When and Why OSHA Inspects Excavation Sites

The Occupational Safety and Health Act of 1970 seeks to "..assure so far as possible every working man and woman in the Nation safe and healthful working conditions."

As one way to promote worker protection, the act authorizes OSHA to set and enforce safety and health standards. The agency conducts inspections to make sure these specific standards are met and that the workplace is generally free from recognized hazards likely to cause death or serious physical harm.

OSHA calls its inspectors compliance safety and health officers (CSHOs). They are experienced professionals whose goal is to help employers and workers reduce on-the-job hazards. While not all of the six million workplaces covered by federal and state OSHAs can be inspected regularly, two basic types of inspections are commonly encountered.

Programmed Inspections

OSHA routinely conducts safety and health inspections in high hazard industries like manufacturing or construction. The agency develops its general schedule for inspecting the most hazardous industries based on various statistical data such as job injury/illness rates, worker compensation, and other information. After entering a workplace in a high hazard business, OSHA inspectors consult and verify the log of injuries and illnesses that most employers with more than ten employees are required to keep. About five percent of OSHA programmed inspections focus on firms in low hazard manufacturing industries, and an additional five percent in non-manufacturing industries.

Follow-up Inspections

The agency (OSHA) may re-inspect firms cited for imminent danger conditions or for willful, repeat, or serious violations to ensure the correction of cited hazards.

OSHA may also conduct follow-up inspections to check the progress of long-term hazard correction programs instituted by employers.

Imminent Danger

OSHA makes inspections of imminent danger situations a top priority. An imminent danger is a hazard that could cause death or serious physical harm immediately, or before the danger could be eliminated through normal enforcement procedures. When compliance officers find imminent danger conditions, they will ask for immediate voluntary correction of the hazard by the employer or removal of endangered employees from the area. If an employer fails to do so, OSHA can go to the nearest federal district court for appropriate legal action.

Fatal Accidents

OSHA also gives high priority to investigating job fatalities and accidents that hospitalize five or more employees. Such accidents must be reported to OSHA within 24 hours. The following examples of vehicle accidents investigated by OSHA illustrate how seemingly innocent construction activities can have deadly consequences.

- An employee was operating a bulldozer at the top edge of a sloped drainage ditch. The bulldozer began to slide down the side of the snow and ice covered excavation, tipped over on its side, and pinned the operator under the roll bars. The driver was not wearing a seat belt.
- A contractor was operating a backhoe when an employee attempted to walk between the swinging superstructure of the backhoe and a concrete wall. As the employee approached from the operator's blind side, the superstructure hit the victim, crushing him against the wall. Employees had not been trained in safe work practices, and no barricades had been erected to prevent employee access to hazardous areas.
- A worker was driving a front-end loader up a dirt ramp onto a lowboy trailer. The tractor was not centered and the tread slipped off the trailer. When the tractor began to tip, the operator jumped from the cab. As he hit the ground, the tractor's rollover protective structure fell on top of him, crushing him. The tractor was not equipped with seat belts.
- While acting as a flagger on a highway construction project, an employee was struck and killed by an oncoming vehicle that failed to stop after being directed to do so.

Complaints

OSHA investigates written and signed complaints made by current employees or their representatives of hazards that threaten serious physical harm to workers. Complaints, other than those involving imminent danger, received from anyone other than a current employee or employee representative, unsigned by a current employee, or received anonymously may result in a letter from the agency to the employer describing the allegation(s) and requesting a response. OSHA will not reveal the name of the person filing the complaint, if so requested.

Chapter 1

Planning for Safe Excavations

Many on-the-job accidents are a direct result of inadequate initial planning. Correcting mistakes in shoring and/or sloping after work has begun slows the operation, adds to its cost, and increases the possibility of an excavation failure. Employers or contractors should build safety into the pre-bid planning in the same way that all other pre-bid factors are considered. The safety of the workers should come first when designing a job operation. A hazard analysis should be completed for all aspects of the job before employees are allowed to begin work on the job.

From the standpoint of planning, contractors must develop a safety checklist before preparing a bid to make certain that there is adequate information about the jobsite and that all necessary items are on hand. In order to ensure safe operations, contractors should incorporate elements of the relevant OSHA standards as well as an understanding of the site characteristics into these checklists.

Before preparing a bid, contractors should take into account these specific site conditions:

- Traffic
- Location of structures and their condition
- Soil
- Surface and ground water
- Water table
- Overhead and underground utilities
- Weather

The design of excavation and trenching is not simple, as soil stability can change with changing weather. Changing weather conditions and climate can also greatly affect how strong a shoring system must be. Excess water from rain or melting snow can loosen the soil and drastically increase the pressure on the shoring system. A rainstorm can turn a stable trench side that required only light bracing into a mass of loose soil, posing an immediate threat to employees working within the excavation. Shoring frozen ground presents another potential problem as a sudden thaw can undermine an entire section of shoring.

Large excavations, in particular, are subject to changing weather conditions because they are generally open for longer periods of time than are trenches. Shoring for these excavations should provide long-range protection for a variety of weather conditions. In some cases, additional precautions are necessary to protect excavation sides or faces, such as covering them with plastic sheeting or spraying the soil bank with a moisture-limiting chemical.

These and other conditions can be determined by jobsite studies, observations, test borings for soil type or conditions, and consultations with local officials and utility companies.

Before any excavation actually begins, the standard requires the employer to determine the estimated location of utility installations—sewer, telephone, fuel, electric, water lines, or any other underground installations—that may be encountered during digging. Also, before starting the excavation, the contractor must contact the utility companies or owners involved and inform them, within established or customary local response times, of the proposed work. The contractor must also ask the utility companies or owners to find the exact location of the underground installations. If they cannot respond within 24 hours (or longer, if required by state or local law) or if they cannot find the exact location of the utility installations the contractor may proceed with caution.

The American Public Works Association (APWA) has available an Excavator Damage Prevention Guide and One-Call Systems Directory. The purpose of One-Call Systems International is to promote facility damage prevention and infrastructure protection through education, guidance, and assistance. The guide is intended as a resource to aid in the prevention of damage to underground facilities. It includes the APWA color code and marking guidelines, a listing of the known international one-call centers, and a summary of damage prevention laws.

One-call systems are about coordination, cooperation, and communication. A one-call system is a tool used to prevent facility dig-ups. While the tool is a useful communication link between excavators and buried-plant owners and operators, it is the responsibility of the excavator to notify the one-call center before excavations occur.

One-call is a safety program designed to cut the cost of pipe and cable repairs and diminish the potential hazard posed to workers and the general public when excavations are done.

Figure 1.1. Located Underground Utilities

This guide can be used as a training and information tool. The one-call center nearest you can also serve as a valuable resource for your training and safety programs; and their phone numbers are listed in the One-Call Systems Directory. Most one-call centers have people available for presentations and have an excellent selection of promotional materials. A copy of the APWA *Excavator's Damage Prevention Guide* is located in Appendix E.

Workers must use safe and acceptable means to find the exact location of underground installations. If underground installations are exposed, OSHA regulations also require that they be removed, protected, or properly supported. An example of located underground utilities is shown in Figure 1.1.

When the necessary specific information about the jobsite is assembled, the contractor is ready to determine the amount, kind, and cost of safety equipment needed. The contractor should make a careful inventory of the safety items on hand before deciding what additional safety material must be acquired. No matter how many trenching, shoring, and backfilling jobs have been done in the past, each job should be approached with the utmost care and preparation.

Have Your Safety Program in Place Before Starting the Job

Before beginning the job, it is important for the contractor to establish and maintain a safety and health program for the worksite that provides adequate systematic policies, procedures, and practices to protect employees from, and allow them to recognize, job-related safety and health hazards.

An effective program includes provisions for the systematic identification, evaluation, and prevention or control of general workplace hazards, specific job hazards, and potential hazards that may arise from foreseeable conditions. The program may be written or verbal but it should reflect the unique characteristics of the jobsite. No attempt should be made to conduct a trenching and shoring operation as part of on-the-job training. Independent contractors must recognize their knowledge limitations and seek advice, assistance, consultation, and specific training as necessary.

Contractors should be aware of the increased potential for excavation collapse due to adverse environmental factors such as elevated levels of ground water. Many incidents occur during the spring "break-up" when large quantities of ice and snow rapidly melt. This phenomenon adds considerable amounts of free water to soil throughout areas of similar geographic and climatic conditions. Soils containing high percentages of silt and gravel can hold significant amounts of groundwater. Independent contractors must be aware of the increased hazards associated with excavation during seasonal periods of higher groundwater levels. Water accumulation in trenches and sloughing off of soil must be carefully observed and are strong indicators of an imminent and immediate danger (see Figure 1.2).

Management's Role in Ensuring Safety

To ensure that safety policies are implemented effectively, it is critical that top management buys into the importance of safety, requires supervisors to be trained in safety techniques, and provides safety training themselves. Each supervisor must understand the degree of responsibility and authority he or she holds in a particular area. For effective labor support, affected unions should be notified in advance of construction plans and asked to cooperate.

Larger and more complex operations should employ a full-time safety official who will recommend ways to improve implementation of the safety plan. In a smaller operation, the safety official may be part-time and usually will be a supervisor.

Supervisors are the contractor's representatives on the job. Supervisors should conduct inspections, investigate accidents, and anticipate hazards. They should ensure that employees receive on-the-job safety and health training. They should also review and strengthen overall safety and health precautions to guard against potential hazards, get the necessary worker cooperation in safety matters, and make frequent reports to the contractor.

Figure 1.2. Water Accumulation

It is important that managers and supervisors set the example for safety at the jobsite. When visiting the jobsite, all managers, regardless of status, must wear the prescribed personal protective equipment such as safety shoes, safety glasses, hard hats, and other necessary gear as required by CFR 1926.100 and 102. CFR 1926.100 requires that employees who work in areas where there is a possible danger of head injury from impact, falling or flying objects, or electrical shock and burns be protected by protective helmets. CFR 1926.102 requires that employees be provided with eye and face protection equipment when machines or operations present potential eye or face injury from physical, chemical, or radiation agents.

The employer should make certain that workers have been properly trained in the use and fit of the prescribed protective gear and equipment, that they are wearing and using the equipment correctly, and that they are following safe work practices. In addition, employees must take an active role in job safety. While management is directly accountable for preventing occupational injuries and illnesses, every employee on the job is responsible for safety.

Chapter 2

Soil Characteristics and Cave-ins

Soil Characteristics and OSHA Standards

When a trench or other excavation is made in soil, the residual forces in the ground work to restore the soil to a more stable configuration. If those residual forces (gravity) are greater than those holding the trench or excavation walls where they are, a cave-in occurs. There are a number of factors that determine the stability of a given excavation wall. It is beyond the scope of this chapter to fully describe soil mechanics in this way. However, it has been found empirically that soil, when sloped appropriately, will resist the residual forces and remain safely stable.

Because the evaluation of soil conditions and structure is crucial to safe operation in and around excavations, an excavation standard was among the first promulgated by OSHA in 1971. In that standard, soils were classified into three types: running, unstable, and hard compact. These terms were generally misunderstood. Later, the terms were revised and renamed granular, cohesive, granular cohesionless, and cemented.

In 1989, it was estimated that 70 fatalities and more than 800 lost workday injuries occurred annually in the United States due to excavation accidents. Responding to this high incidence rate, OSHA promulgated the current excavation standards (Ref 3.1), which has the following requirements:

1. **Classification of soil and rock deposits.** Each soil and rock deposit shall be classified by a competent person as Stable Rock, Type A, Type B, or Type C in accordance with the definitions set forth in 29 CFR 1926 Subpart P Appendix A.
2. **Basis of classification.** The classification of the deposits shall be made based on the results of at least one visual and at least one manual analysis. Such analyses shall be conducted by a competent person using tests described in 29 CFR 1926

Subpart P Appendix A, or using other recognized methods of soil classification and testing such as those adopted by the American Society for Testing Materials, or the U.S. Department of Agriculture textural classification system.

3. **Visual and manual analyses.** The visual and manual analyses, such as those noted as being acceptable in 29 CFR 1926 Subpart P Appendix A, shall be designed and conducted to provide sufficient quantitative and qualitative information as may be necessary to properly identify the properties, factors, and conditions affecting the classification of the deposits.
4. **Layered systems.** In a layered system, the system shall be classified in accordance with its weakest layer. However, each layer may be classified individually where a more stable layer lies under a less stable layer.
5. **Reclassification.** If, after classifying a deposit, the properties, factors, or conditions affecting its classification change in any way, the changes shall be evaluated by a competent person. The deposit shall be reclassified as necessary to reflect the changed circumstances.

The angle of the slope depends upon the properties of the soil in which the excavation has been made. The mineral particles that constitute soil and other earth materials can bond by chemical and physical forces that oppose the force of gravity. Chemical bonding, or cohesion, refers to the chemical forces that bond mineral particles. Physical bonding refers to the bonding of the more coarse grains such as sand and gravel by frictional forces that include the interlocking of particles.

29 CFR 1926 Subpart P Appendix A is based on site and environmental conditions, and on the structure and composition of the soil deposits. The soil classification system is a method of categorizing soil and rock deposits in a hierarchy of Stable Rock, Type A, Type B, and Type C, in decreasing order of stability.

1. Stable rock is a natural solid mineral matter that can be excavated with vertical sides and remain intact while exposed.
2. Type A soils are cohesive soils with an unconfined, compressive strength of 1.5 ton per square foot (tsf) or greater. Examples of cohesive soils are: clay, silty clay, sandy clay, clay loam, and in some cases, silty clay loam and sandy clay loam. Cemented soils such as caliche and hardpan are also considered Type A. However, no soil is Type A if:
    a. The soil is fissured.
    b. The soil is subject to vibration from heavy traffic, pile driving, or similar effects.
    c. The soil has been previously disturbed.
    d. The soil is part of a sloped, layered system where the layers dip into the excavation on a slope of four horizontal to one vertical (4H:1V) or greater.

   e. The material is subject to other factors that would require it to be classified as a less stable material.

3. Soils classified as Type B can be:
   a. Cohesive soil with an unconfined compressive strength greater than 0.5 tsf but less than 1.5 tsf.
   b. Granular cohesionless soils including: angular gravel (similar to crushed rock), silt, silt loam, sandy loam, and in some cases, silty clay loam and sandy clay loam.
   c. Previously disturbed soils except those that would otherwise be classed as Type C soil.
   d. Soil that meets the unconfined compressive strength or cementation requirements for Type A, but is fissured or subject to vibration.
   e. Dry rock that is not stable.
   f. Material that is part of a sloped, layered system where the layers dip into the excavation on a slope less steep than four horizontal to one vertical (4H:1V), but only if the material would otherwise be classified as Type B.
4. Type C includes:
   a. Cohesive soil with an unconfined compressive strength of 0.5 tsf or less.
   b. Granular soils including gravel, sand, and loamy sand.
   c. Submerged soil or soil from which water is freely seeping.
   d. Submerged rock that is not stable.
   e. Material in a sloped, layered system where the layers dip into the excavation or a slope of 4H:1V or steeper.

OSHA Method ID-194 was developed to emphasize the performance and engineering properties of soil and is consistent with the objectives and requirements of the 1989 federal excavation regulations. Many other methods used to classify soil provide textural names for soil such as clay, silty clay, sandy clay, clay loam, and in some cases, silty clay loam and sandy clay loam. Because these other methods specify slightly different tests, the names used to classify soil can be similar but have different meanings. OSHA attempted to provide a textural name of the soil with their method (so that a comparison could be made with other methods), but a complete separation of all of the soil components is not performed by the tests described in this method. This could result in a textural name that is slightly different from those used in other soil classifying methods.

Cave-ins

Trenching and excavation work presents serious risks to all workers involved. Of the many hazards to which excavation workers are exposed, the danger of cave-ins is the greatest. When cave-in accidents occur, they are more likely to result in worker fatalities than other excavation-related accidents (see Figure 2-1). Strict compliance, however, with all sections of the standard will prevent or greatly reduce the risk of cave-ins as well as other excavation-related accidents.

Figure 2.1. Potential Cave-in.

Definition of a Cave-in

A cave-in is defined by OSHA as "The separation of a mass of soil or rock material from the side of an excavation, or the loss of soil from under a trench shield or support system, and its sudden movement into the excavation, either by falling or sliding, in sufficient quantity so that it could entrap, bury, or otherwise injure and immobilize a person". OSHA requires that in all excavations employees exposed to potential cave-ins must be protected by sloping, benching the sides of the excavation, supporting the sides of the excavation, or placing a shield between the side of the excavation and the work area.

Why Do Cave-ins Occur?

Cave-ins occur because an open excavation is an unnatural situation. If you look out the window, your average, natural landscape shows no vertical or near vertical slopes. Soil or dirt is a very heavy material. One cubic foot can easily weigh 114 pounds and 1 cubic yard can be as heavy as a pick-up truck.

To get a sense of the pressures involved in excavation work, consider a square column of soil that measures 1 foot by 1 foot across and several feet high. At 1 foot of depth, a horizontal plane 1 foot by 1 foot carries the cubic foot that lies over it. The stress, or load per unit area, is 100 pounds per square foot (psf). At a depth of 2 feet, the horizontal plane carries 2 cubic feet or 200 psf. At a depth of 5 feet, the vertical stress is 500 psf, and so on.

The column described would soon collapse under the stress if not supported by similar adjacent columns. Adjacent columns supply stresses that act horizontally on

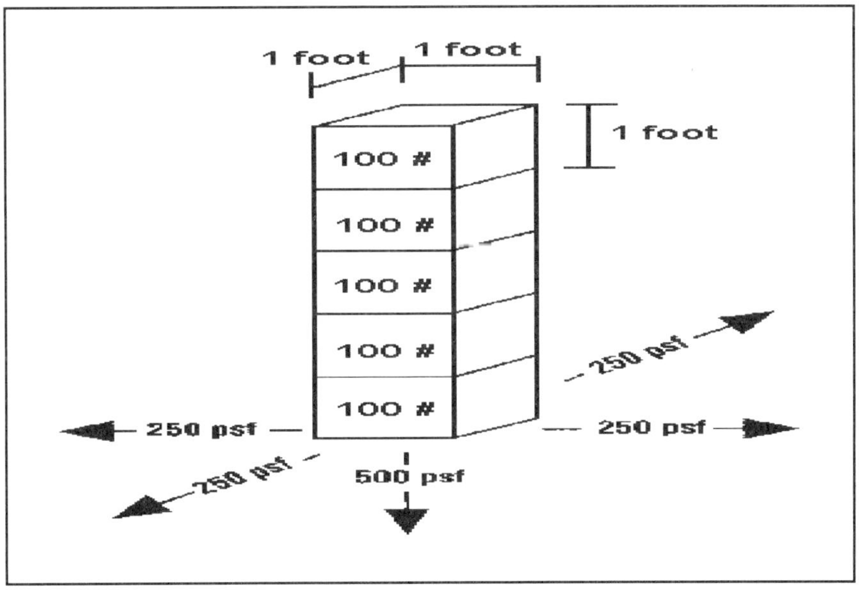

Figure 2.2. How Heavy Is Soil?

the column. However, these lateral stresses can be considered half as large as the vertical stresses. Figure 2-2 shows that at a depth of 5 feet, the vertical stress is 500 psf and the lateral, or horizontal stress, is 250 psf.

Undisturbed soil may be visualized as an infinite number of columns of soil adjoining and supporting one another. The system is in equilibrium and is perfectly stable.

When an excavation is cut, the system is disturbed. Lateral stresses that existed on the excavation wall are removed as the excavation is done. The soil in the excavation wall immediately begins to move, however slowly, into the excavation.

At the same time, the surface of the ground next to the excavation subsides, creating an unnatural situation. The surface of the ground is in tension and some of the weight of the soil in the excavation wall is transferred to the soil back away from the wall face by a phenomenon called shear.

The combination of tension in the ground surface and shear stress causes cracks to form back from the edge of the excavation. Cracks occur from approximately the depth of the excavation back from its edge. For example, if there is an excavation 10 feet deep, the cracks may be found somewhere between 3 to 7 feet back from the excavation edge. There may be several cracks. They are usually vertical and they may be half the depth of the excavation.

When cracks develop, the weight of the soil in the excavation wall is no longer partly carried by the soil back from the excavation's face. The lower part of the excavation wall fails under the great stress caused by the weight of the soil above the wall. There is no lateral stress to prevent the failure.

When the bottom of the excavation fails, or "kicks," into the excavation, the support for the upper part of the excavation wall is now essentially hanging only by shear and tension forces and failure occurs.

A third cave-in quickly follows. Soil, like concrete, is normally strong in compression but not at all strong in tension. Reinforced concrete makes use of the compressive strength of concrete and the tensile strength of steel. There is no steel in the soil.

Cave-ins generally come in multiples. If the first one doesn't get you, the second one may, and there is always the possibility of a third.

Unstable Soil Conditions

There are several soil conditions that signal dangerous situations and indicate that a cave-in is possible. Some of these conditions are illustrated in Figure 2-3.

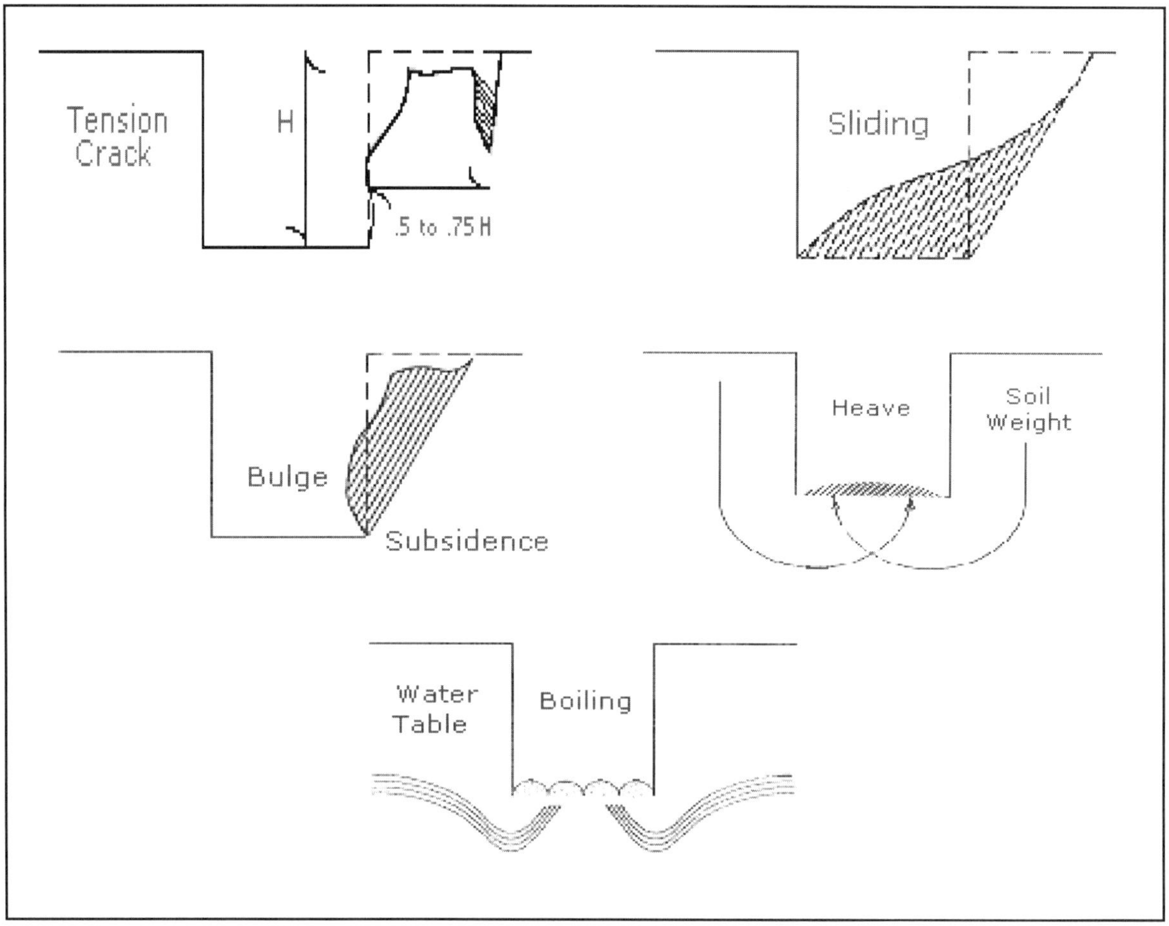

Figure 2.3. Unsafe Soil Conditions.

Tension cracks—Tension cracks usually form at a horizontal distance of 0.5 to 0.75 times the depth of the trench, measured from the top of the vertical face of the trench.

Sliding or sluffing—Sliding or sluffing may occur as a result of tension cracks.

Toppling—In addition to sliding, tension cracks can cause toppling. Toppling occurs when the trench's vertical face shears along the tension crack line and topples into the excavation.

Subsidence and bulging—An unsupported excavation can create an unbalanced stress in the soil, which, in turn, causes subsidence at the surface and bulging of the vertical face of the trench. If uncorrected, this condition can cause face failure and entrapment of workers in the trench.

Heaving or squeezing—The downward pressure created by the weight of adjoining soil causes bottom heaving or squeezing. This pressure causes a bulge in the bottom of the cut, as illustrated in the drawing above. Heaving and squeezing can occur even when shoring or shielding has been properly installed.

Boiling—Boiling is evidenced by an upward water flow into the bottom of the cut. A high water table is one of the causes of boiling. Boiling produces a "quick" condition in the bottom of the cut, and can occur even when shoring or trench boxes are used.

This explanation of the mechanics of a cave-in has offered a discussion of some of the forces involved in such accidents. It has by no means considered all of the forces that may be involved in such an occurrence. Weathering, water, vibration, and superimposed loads may add to the hazardous conditions leading to cave-ins.

Summary

Cave-ins during excavations result in several fatalities annually in the United States. A cave-in as defined by OSHA is "the separation of a mass of soil or rock material from the side of an excavation, or the loss of soil from under a trench shield or support system, and its sudden movement into the excavation, either by falling or sliding, in sufficient quantity so that it could entrap, bury, or otherwise injure and immobilize a person."

Undisturbed soil is kept in place by the horizontal and vertical forces of adjacent soil. The soil removed to create a trench is no longer available to provide support for the soil left behind in the trench or excavation wall. With no support, soil from the wall will eventually move downward and inward into the excavation. This creates a life-threatening hazard for workers in the excavation. No one can predict accurately if an excavation is safe to enter without a support structure present.

A cube of soil measuring 1 foot on each side weighs approximately 100 pounds. A cubic yard of soil contains 27 of these or around 2700 pounds, about as much as a small automobile. A trench or excavation wall could contain 3 to 5 cubic yards of soil, weighing from 8,000 to 14,000 pounds. With little or no warning, an unsupported, improperly shored or sloped trench or excavation wall can collapse suddenly and with great force, burying workers in seconds beneath tons of soil and debris before they have a chance to react. A worker does not have to be completely buried in soil to be seriously injured or killed.

Chapter 3

Protecting against Excavation Hazards

Support Systems to Protect against Cave-ins

Designing a protective system can be a complex process because of the number of factors involved—soil classification, depth of cut, water content of soil, changes due to weather and climate, or other operations in the vicinity. The standard, however, provides several different methods and approaches.

One approach involves designing protective systems that can be used to provide the level of protection needed to prevent cave-ins.

Sloping

One method of ensuring the safety and health of workers in an excavation is to slope the sides of an excavation (see Figure 3.1).

Figure 3.1. Sloping

The steepness of the slope (or *angle*) depends on the type of soil that is used. Table 3.1 gives the allowable angles for soil types A, B, and C.

Table 3.1. Table of Sloping

Soil Type	Height/ Width Ratio	Slope Angle (deg.)	Depth (feet)	Total Distance (feet)
A	3/4:1	53	10	15 + bottom width
B	1:1	45	10	20 + bottom width
C	11/2:1	34	10	30 + bottom width

One and One-Half Horizontal to Vertical

A slope not steeper than one and one-half horizontal to one vertical (34 degrees measured from the horizontal) is considered safe for any type of soil. These slopes must be excavated to form configurations that are in accordance with those for Type C soil found in Appendix B of the standard.

All simple slope excavations 20 feet or less in depth shall have a maximum allowable slope of 1 1/2:1.

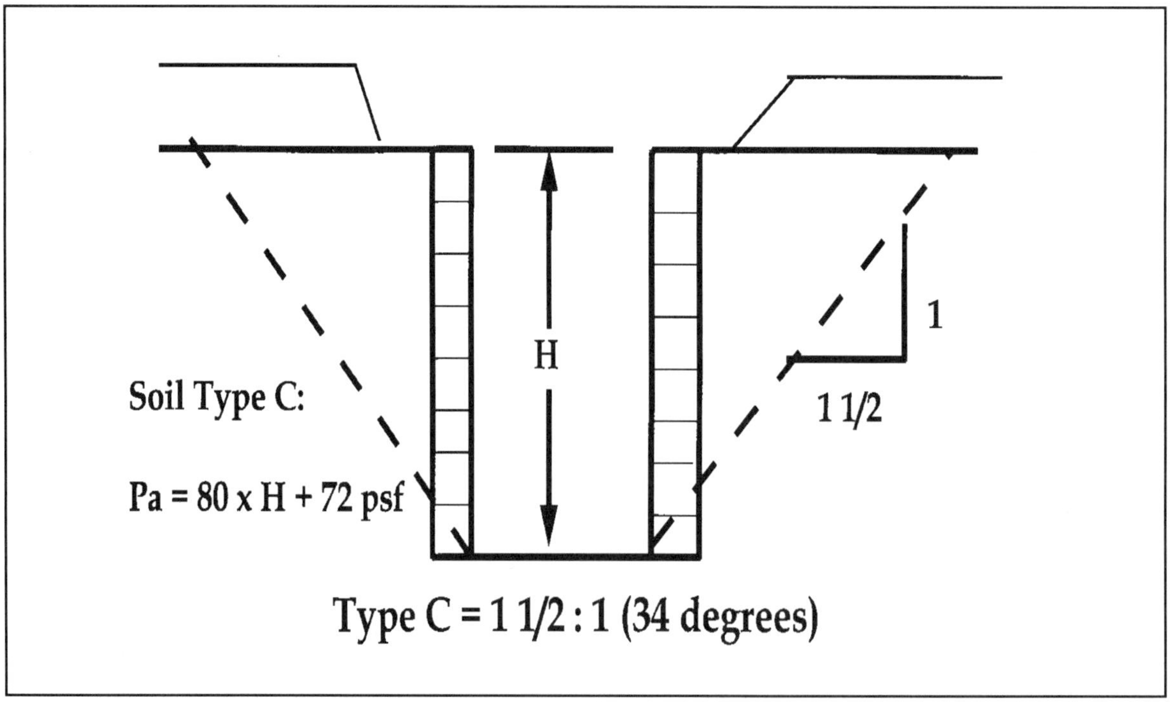

Figure 3.2. Sloping Design: Type C Soil.

Use of Tabulated Data

A second design method, which can be applied for both sloping and shoring, involves using tabulated data, such as tables and charts, that have been approved by a registered professional engineer. These data must be in writing and must include sufficient explanatory information to enable the user to make a selection, including the criteria for determining the selection and the limits on the use of the data.

At least one copy of the information, including the identity of the registered professional engineer who approved the data, must be kept at the worksite during construction of the protective system. Upon completion of the system, the data may be stored away from the jobsite, but a copy must be made available, upon request, to the Assistant Secretary of Labor for OSHA.

Trench Boxes and Shields

Contractors also may use a trench box, or shield, that is either designed or approved by a registered professional engineer or is based on tabulated data prepared or approved by a registered professional engineer. Timber, aluminum, or other suitable materials may also be used. OSHA standards permit the use of a trench shield (also known as a welder's hut) as long as the protection it provides is equal to or greater than the protection that would be provided by the appropriate shoring system (see Figures 3.3 and 3.4).

The employer is free to choose the most practical design approach for any particular circumstance. Once an approach has been selected, however, the required performance criteria must be met by that system.

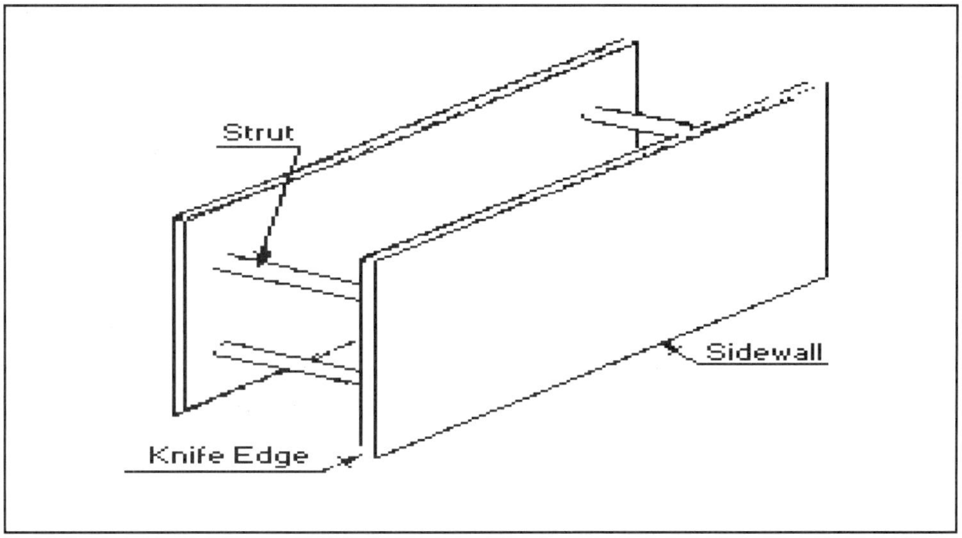

Figure 3.3. Trench Box.

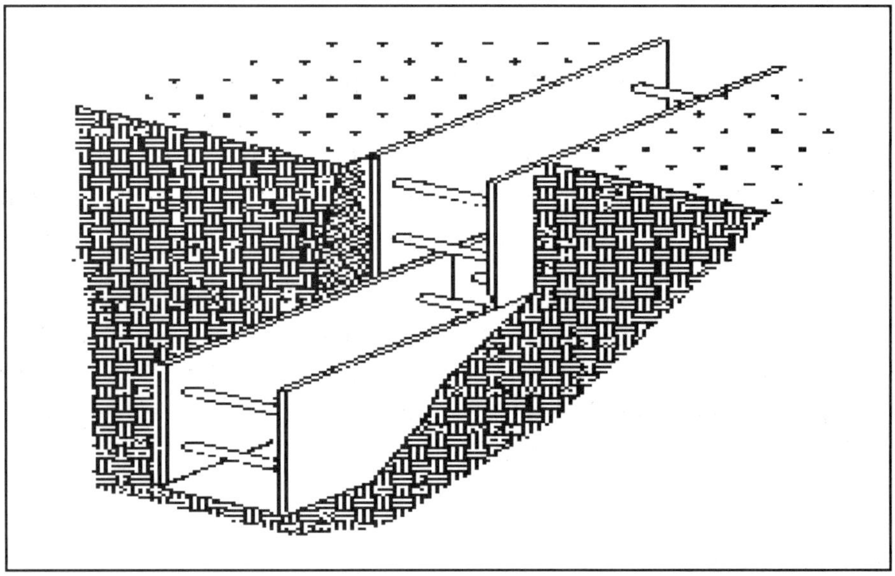

Figure 3.4. Shielding.

Excavating near Buildings, Walls, and Sidewalks

When excavating adjacent to structures such as buildings, walls, sidewalks, pavements, or other structures, the standard also requires the employer to provide support systems such as shoring, bracing, or underpinning to ensure the stability of the adjacent structures (see Figure 3.5).

In addition, the standard prohibits excavation below the level of the base or footing of any foundation or retaining wall unless (1) a support system such as underpinning is provided, (2) the excavation is in stable rock, or (3) a registered professional engineer determines that the structure is sufficiently removed from the excavation and that excavation will not pose a hazard to employees.

Excavations under sidewalks and pavements are also prohibited unless an appropriately designed support system is provided or another effective method is used.

Exceptions to the Standard

The standard does not require the installation and use of a protective system when an excavation (1) is made entirely in stable rock, or (2) is less than 5 feet deep and a competent person has examined the ground and found no indication of a potential cave-in.

Figure 3.5. Support System for Adjacent Structure.

Installation and Removal of Protective Systems

The standard requires that during the installation of support systems, these procedures be followed to protect employees:

- Securely connect members of support systems.
- Safely install support systems.
- Never overload members of support systems.
- Install other structural members to carry loads imposed on the support system when temporary removal of individual members is necessary.

In addition, the standard permits excavation of 2 feet or less below the bottom of the members of a support or shield system of a trench if (1) the system is designed to resist the forces calculated for the full depth of the trench, and (2) there are no indications while the trench is open of a possible cave-in below the bottom of the support system. Also, the installation of support systems must be closely coordinated with the excavation of trenches.

As soon as work is completed, the excavation should be back-filled as the protective system is dismantled. After the excavation has been cleared, workers should slowly remove the protective system from the bottom up, taking care to release members slowly.

Safe Access/Egress

Under the standard, the employer must provide safe access and egress to all excavations. According to OSHA regulations, when employees are required to be in trench excavations 4-feet deep or more, adequate means of exit, such as ladders, steps, ramps, or other safe means of egress, must be provided and be within 25 feet of lateral travel. As stated in the following citation, a competent person must design the structural ramps used as a means of employee or vehicle access or egress.

"Structural ramps that are used solely by employees as a means of access or egress from excavations shall be designed by a competent person. Structural ramps used for access or egress of equipment shall be designed by a competent person qualified in structural design, and shall be constructed in accordance with the design." [1926.651(c)(1)(i)].

Also, structural members used for ramps or runways must be uniform in thickness and joined in a manner to prevent tripping or displacement.

Dangerous Materials and Equipment

The employer is responsible for the safe condition of materials and equipment used for protective systems. Defective and damaged materials and equipment can result in the failure of a protective system and create excavation hazards.

To avoid possible failure of a protective system, the employer must ensure that:

(1) Materials and equipment are free from damage or defects.

(2) Manufactured materials and equipment are used and maintained in a manner consistent with the recommendations of the manufacturer and in a way that will prevent employee exposure to hazards.

(3) While in operation, damaged materials and equipment are examined by a competent person to determine if they are suitable for continued use. If materials and equipment are not safe for use, they must be removed from service. These materials cannot be returned to service without the evaluation and approval of a registered professional engineer.

Falls, Falling Loads, and Mobile Equipment

In addition to cave-in hazards and secondary hazards related to cave-ins, there are other hazards from which workers must be protected during excavation-related work. These hazards include exposure to falls, falling loads, and mobile equipment. To protect employees from these hazards, OSHA requires the employer to take the following precautions:

- Keep materials or equipment that might fall or roll into an excavation at least 2 feet from the edge of excavations, ensure that they have retaining devices, or both.
- Provide warning systems, such as mobile equipment, barricades, hand or mechanical signals, or stop logs, to alert operators of the edge of an excavation. When possible, the soil should be graded away from the excavation; this will assist in vehicle control and channeling of run-off water.
- Provide scaling to remove loose rock or soil or install protective barricades and other equivalent protection to protect employees from falling rock, soil, or other materials.
- Prohibit employees from working on faces of sloped or benched excavations at levels above other employees, unless employees at lower levels are adequately protected from the hazard of falling, rolling, or sliding material or equipment.
- Prohibit employees under loads that are handled by lifting or digging equipment. To avoid being struck by any spillage or falling materials, require employees to stand away from vehicles being loaded or unloaded. If cabs of vehicles provide adequate protection from falling loads during loading and unloading operations, the operators may remain in them.

Barricades

Barricades are frequently used to protect employees from dangerous equipment at construction sites. For example, a barricade around the rotating superstructure of a crane prevents an employee from being struck by the superstructure or crushed between the superstructure and other objects or another part of the crane. The hazard is the same when the superstructure of excavating equipment is rotated.

When the same basic machines can be used for either lifting or excavating, the safety precautions required for cranes should logically be applied to excavating equipment. Since it is common practice to provide barricades around the swing radius of the superstructure of the machine when it is used as a crane, it should be common practice to provide such a barricade when the machine is used as a shovel or backhoe.

For this reason, it is OSHA policy to apply 1926.550(a)(9) to excavating equipment when the same basic machine can be used for either lifting or excavating. Where a backhoe is digging below or near powerlines, insulating barriers are adequate only if the reasonably expected motion of a backhoe boom and bucket could produce, at most, a brushing type contact with an overhead wire and damage to the wire or barrier would not be possible or expected. If, on the other hand, the reasonably expected motion of a backhoe could pull down or contact conductors or otherwise damage the wire or barrier if contact was made, then barriers would not be adequate. Under these conditions, the powerline would have to be deenergized or the backhoe moved away from the lines.

Water Accumulation

The standard prohibits employees from working in excavations where water has accumulated or is accumulating unless adequate protection has been taken (see Figure 1.2). If water removal equipment is used to control or prevent water from accumulating, a competent person must monitor the equipment and operations of the equipment to ensure proper use.

OSHA standards also require that diversion ditches, dikes, or other suitable means be used to prevent surface water from entering an excavation and to provide adequate drainage of the area adjacent to the excavation. Also, a competent person must inspect excavations subject to runoffs from heavy rains.

Hazardous Atmospheres

Under this provision, a competent person must test excavations greater than 4 feet in depth as well as excavations where oxygen deficiency or a hazardous atmosphere exists or could reasonably be expected to exist, before an employee enters the excavation. If hazardous conditions exist, controls such as proper respiratory protection or ventilation must be provided. Also, controls used to reduce atmospheric contaminants to acceptable levels must be tested regularly. The frequency of testing should be increased if equipment is operating in the trench. Testing frequency should also be increased if welding, cutting, or burning is done in the trench.

Where adverse atmospheric conditions may exist or develop in an excavation, the employer also must provide emergency rescue equipment, (e.g., breathing apparatus, a safety harness and line, basket stretcher, etc.) and ensure that it is readily available. This equipment must be attended when used.

When an employee enters bell-bottom pier holes and similar deep and confined footing excavations, the employee must wear a harness with a lifeline. The lifeline must be securely attached to the harness and must be separate from any line used to handle materials. Also, while the employee wearing the lifeline is in the excavation, an observer must be present to ensure that the lifeline is working properly and to maintain communication with the employee.

Summary

The greatest hazard to workers is working in an excavation that is not protected or inadequately protected from cave-ins. The following is an analysis of the number of serious excavation violations for the year 1996. This shows that the largest numbers of violations were related to the lack of protective systems in use.

Figure 3.6. Hazardous Atmosphere.

In Figure 3.7, the category axis on the left is the OSHA standard cited violation while the bar on the right provides the number of violations for the year.

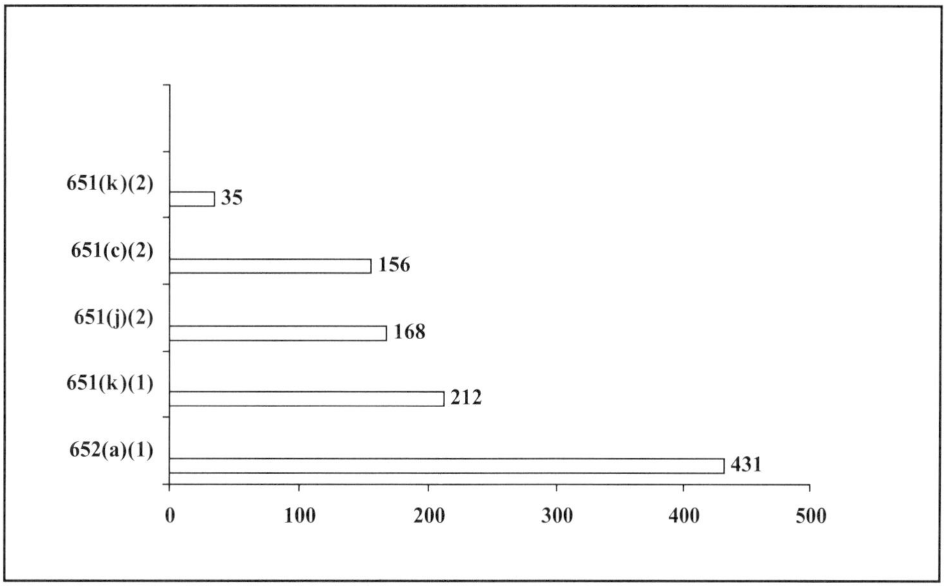

Figure 3.7. OSHA Standard Cited Violations

Strict compliance with OSHA standards will help ensure a good level of safety and health on the job. However, what makes workplace safety and health a reality is a diligent and committed workforce operating as a team, dedicated to the principle that safety is the responsibility of every employee and all occupational injuries and illnesses can be prevented.

To prevent trench cave-ins and additional worker fatalities, OSHA requires that one or more of the following precautions must be taken when working in excavations:

1. Utilize a shield or trench box system designed to protect workers in excavations.
2. Shore sides of excavations with timber or other materials to ensure that the earth does not collapse on workers who must enter them.
3. Slope the sides of excavations to reduce the "overburden" (weight and pressure exerted by large amounts of soil on the sides).
4. Secure sides by equivalent means, such as engineer-designed sheeting or bracing.
5. Assure that the sides of all excavations are shored, laid back to a stable slope, or provided with other equivalent protection where employees may be exposed to moving ground or cave-ins.
6. Have a competent person frequently inspect excavations where the soil was previously disturbed or where there are variations in the moisture content of the soil, loading due to equipment or stored materials, or vibration from equipment or traffic.
7. Train employees, including periodic refresher training, to be aware of and understand the hazards of the job.
8. Perform an initial hazard assessment of the job prior to beginning work and when a situation change presents different hazards.
9. In cases where employees not associated with the excavation work may have to approach construction areas, a safe path of entry should be planned with physical barriers and/or warning lines that mark the boundaries of the construction area.

Excavation fatalities and injuries must not be accepted as an inevitable annual occurrence. Caution must be taken whenever workers are exposed to life-threatening hazards. Injuries and deaths can be prevented if employers and employees are aware of the hazards they face and ensure that proper precautions are taken.

Chapter 4

Safety Training

OSHA Requirements

Many standards promulgated by the Occupational Safety and Health Administration explicitly require the employer to train employees in the safety and health aspects of their jobs. Other OSHA standards make it the employer's responsibility to limit certain job assignments to employees who are "certified," "competent," or "qualified," meaning that they have had special previous training in or out of the workplace. The term "designated" applies to personnel selected or assigned by the employer or the employer's representative as qualified to perform specific duties. These requirements reflect OSHA's belief that training is an essential part of every employer's safety and health program for protecting workers from injuries and illnesses. Many researchers conclude that those who are new on the job have a higher rate of accidents and injuries than more experienced workers. If ignorance of specific job hazards and proper work practices is even partly to blame for this higher injury rate, training will help to provide a solution.

In OSHA Directives CPL 2.87, "Inspection Procedures for Enforcing the Excavation Standard, 29 CFR 1926, Subpart P," OSHA compliance personnel are provided guidance on ensuring compliance with 29 CFR 1926, Subpart P, Excavations. This includes the following:

- An evaluation of the training of affected employees and the effectiveness of the employer's enforcement of its program. (See 29 CFR 1926.20(b)(1) and 1926.21(b)(2).)
- Evaluate compliance with requirements for periodic inspection of excavations. (See 29 CFR 1926.651 (k)(1).)
- Identify all persons (competent person, registered professional engineer, etc.) responsible for excavation activities and/or operations.

- Evaluate compliance with training requirements identified by periodic inspections or changes in equipment and/or procedures.

This shall include an evaluation of the effectiveness of the employer's inspection procedures and training program for assessment and correction of situations resulting in near misses and/or injuries, or circumstances indicating that modifications are necessary. (See 29 CFR 1926.20(b)(1) and 1926.21(b)(2).)

As an example of the trend in OSHA safety and health training requirements, the *Process Safety Management of Highly Hazardous Chemicals Standard (29 CFR Part 1926.64)* contains several training requirements. This standard was promulgated under the requirements of the Clean Air Act Amendments of 1990. The Process Safety Management Standard requires the employer to evaluate or verify that employees comprehend the training given to them. The training program must have established goals and objectives regarding expected accomplishments. Subsequent to the training, evaluations should be conducted to verify that the employees understood the subjects presented or acquired the desired skills. If the established goals and objectives of the training program are not achieved as expected, the employer should revise the training program to make it more effective, conduct more frequent refresher training, or some combination of both. The requirements of the Process Safety Management Standard follow the concepts embodied in the OSHA training guidelines contained in this discussion.

The length and complexity of OSHA standards may make it difficult to find all the references to training. So, to help employers, safety and health professionals, training directors, and others with a need to know, OSHA's training-related requirements have been excerpted and collected in this discussion. Requirements for posting information, warning signs, labels, and the like are excluded, as are most references to the qualifications of people assigned to test workplace conditions or equipment.

It is usually a good idea for the employer to keep a record of all safety and health training. Records can provide evidence of the employer's good faith and compliance with OSHA standards. Documentation can also answer one of the first questions an accident investigator will ask: "Was the injured employee trained to do the job?"

Training in proper job performance is time and money well spent, and the employer might regard it as an investment rather than an expense. An effective program of safety and health training for workers can result in fewer injuries and illnesses, better morale, and lower insurance premiums, among other benefits.

Voluntary Training Guidelines

The Occupational Safety and Health Act of 1970 does not address specifically the responsibility of employers to provide health and safety information and instruction to employees, although Section 5(a)(2) does require that each employer . . . shall comply with occupational safety and health standards promulgated under this Act." However, more than 100 of the act's current standards do contain training requirements.

Therefore, the Occupational Safety and Health Administration has developed voluntary training guidelines to assist employers in providing the safety and health information and instruction needed for their employees to work at minimal risk to themselves, to fellow employees, and to the public.

The guidelines are designed to help employers:

(1) Determine whether a worksite problem can be solved by training.

(2) Determine what training, if any, is needed.

(3) Identify training goals and objectives.

(4) Design learning activities.

(5) Conduct training.

(6) Determine the effectiveness of the training.

(7) Revise the training program based on feedback from employees, supervisors, and others.

The development of the guidelines is part of an agency-wide objective to encourage cooperative, voluntary safety and health activities among OSHA, the business community, and workers. These voluntary programs include training and education, consultation, voluntary protection programs, and abatement assistance.

Training Model

The guidelines provide employers with a model for designing, conducting, evaluating, and revising training programs. The training model can be used to develop training programs for a variety of occupational safety and health hazards identified at the workplace. Additionally, it can assist employers in their efforts to meet the training requirements in current or future occupational safety and health standards.

A training program designed in accordance with these guidelines can be used to supplement and enhance the employer's other education and training activities. The guidelines afford employers significant flexibility in the selection of content and training program design. OSHA encourages a personalized approach to the informational and instructional programs at individual worksites; programs tailored to specific sites enable employers to provide the training that is most needed and applicable to local working conditions.

Assistance with training programs or the identification of resources for training is available through such organizations as OSHA full-service area offices, state agencies that have their own OSHA-approved occupational safety and health programs, OSHA-funded state onsite consultation programs for employers, local safety councils, the OSHA Office of Training and Education, and OSHA-funded New Directions grantees.

Review Commission Implications

OSHA does not intend to make the guidelines mandatory. And employers should not use them as a total or complete guide in training and education matters that can result in enforcement proceedings before the Occupational Safety and Health Review Commission. However, employee training programs are always an issue in Review Commission cases that involve alleged violations of training requirements contained in OSHA standards.

The adequacy of employee training may also become an issue in contested cases where the affirmative defense of unpreventable employee misconduct is raised. Under case law well-established in the commission and the courts, an employer may successfully defend against an otherwise valid citation by demonstrating that all feasible steps were taken to avoid the occurrence of the hazard, and that actions of the employee involved in the violation were a departure from a uniformly and effectively enforced work rule of which the employee had either actual or constructive knowledge.

In either type of case, the adequacy of the training given to employees in connection with a specific hazard is a matter that can be decided only by considering all the facts and circumstances surrounding the alleged violation. The general guidelines in this publication are not intended, and cannot be used, as evidence of the appropriate level of training in litigation involving either the training requirements of OSHA standards or affirmative defenses based upon employer training programs.

The Training Process in Seven Steps

OSHA's training guidelines follow a model that consists of:

(1) Determining if Training Is Needed

(2) Identifying Training Needs

(3) Identifying Goals and Objectives

(4) Developing Learning Activities

(5) Conducting the Training

(6) Evaluating Program Effectiveness

(7) Improving the Program

The model is designed so that even the owner of a business with very few employees can use it without having to hire a professional trainer or purchase expensive training materials. Using this model, employers or supervisors can develop and administer safety and health training programs that address problems specific to their own business, fulfill the learning needs of their own employees, and strengthen the overall safety and health program of the workplace.

Determining if Training Is Needed

The first step in the training process is a basic one to determine whether a specific problem can be resolved by training. Whenever employees are not performing their jobs properly, it is often assumed that training will bring them up to standard. However, it is possible that other actions (such as hazard abatement or the implementation of engineering controls) would enable employees to perform their jobs properly.

Ideally, safety and health training should be provided before problems or accidents occur. This training would cover both general safety and health rules and work procedures, and would be repeated if an accident or near-miss incident occurred.

Problems that can be addressed effectively by training include those that arise from lack of knowledge of a work process, unfamiliarity with equipment, or incorrect execution of a task. Training is less effective (but still can be used) for problems arising from an employee's lack of motivation or lack of attention to the job. Whatever its purpose, training is most effective when designed in relation to the goals of the employer's total safety and health program.

Identifying Training Needs

If the problem can be solved in whole or in part by training, the next step is to determine what training is needed. To determine appropriate training, it is necessary to identify what the employee is expected to do and in what ways, if any, the employee's performance is deficient. This information can be obtained by conducting a job analysis that pinpoints what an employee needs to know in order to perform a job.

When designing a new training program or preparing to instruct an employee in an unfamiliar procedure or system, a job analysis can be developed by examining engineering data on new equipment or the safety data sheets on unfamiliar substances. The content of the specific federal or state OSHA standards applicable to a business can also provide direction in developing training content. Another option is to conduct a Job Hazard Analysis. This is a procedure used to study and record each step of a job, identify existing or potential hazards, and determine the best way to perform the job in order to reduce or eliminate risks. Information obtained from a Job Hazard Analysis can be used as the content for the training activity.

If an employer's learning needs can be met by revising an existing training program rather than developing a new one, or if the employer already has some knowledge of the process or system to be used, appropriate training content can be developed through such means as:

(1) Using company accident and injury records to identify how accidents occur and what can be done to prevent them from recurring.

(2) Requesting employees to provide, in writing and in their own words, descriptions of their jobs. These should include the tasks performed and the tools, materials, and equipment used.

(3) Observing employees at the worksite as they perform tasks, asking about the work, and recording their answers.

(4) Examining similar training programs offered by other companies in the same industry, or obtaining suggestions from such organizations as the National Safety Council (which can provide information on Job Hazard Analysis), the Bureau of Labor Statistics, OSHA-approved state programs, OSHA full-service area offices, OSHA-funded state consultation programs, or the OSHA Office of Training and Education.

The employees themselves can provide valuable information on the training they need. Safety and health hazards can be identified through the employees' responses to questions such as whether anything about their jobs frightens them, if they have had any near-miss incidents, if they feel they are taking risks, or if they believe that their jobs involve hazardous operations or substances.

Once the kind of training needed has been determined, it is equally important to determine the kind of training that is **not** needed. Employees should be made aware of all the steps involved in a task or procedure, but training should focus on those steps where performance must be improved. If this is taken into consideration, unnecessary training is avoided and the training is tailored to meet the needs of the employees.

Identifying Goals and Objectives

Once the employees' training needs have been identified, employers can prepare objectives for the training. Instructional objectives, if clearly stated, will tell employers what they want their employees to do, to do better, or to stop doing.

Learning objectives do not necessarily have to be written, but in order for the training to be as successful as possible, clear and measurable objectives should be thought out before the training begins. For an objective to be effective, it should identify as precisely as possible how the individuals will demonstrate what they have learned or that the objective has been reached. They should also describe the important conditions under which the individual will demonstrate competence and define what constitutes acceptable performance.

Using specific, action-oriented language, the instructional objectives should describe the preferred practice or skill and its observable behavior. For example, rather than using the statement, "The employee will understand how to use a respirator" as an instructional objective, it would be better to say, "The employee will be able to describe how a respirator works and when it should be used." Objectives are most effective when worded in sufficient detail that other qualified persons can recognize when the desired behavior is exhibited.

Developing Learning Activities

Once employers have stated precisely what the objectives for the training program are, learning activities can be identified and described. Learning activities enable employees to demonstrate that they have acquired the desired skills and knowledge. To ensure that employees transfer the skills or knowledge from the learning activity to the job, the learning situation should simulate the actual job as closely as possible. Thus, employers may want to arrange the objectives and activities in a sequence that corresponds to the order in which the tasks are to be performed on the job, if a specific process is to be learned. For instance, if an employee must learn the beginning processes of using a machine, the sequence might begin with:

(1) Check that the power source is connected.

(2) Ensure that the safety devices are in place and are operative.

(3) Know when and how to throw the switch.

A few factors will help to determine the type of learning activity to be incorporated into the training. One aspect is the training resources available to the employer. Can a group training program that uses an outside trainer and film be organized, or should the employer personally train the employees on a one-to-one basis? Another factor is the kind of skills or knowledge to be learned. Is the learning oriented toward physical skills (such as the use of special tools) or toward mental processes and attitudes? Such factors will influence the type of learning activity designed by employers. The training activity can be group-oriented, with lectures, role playing, and demonstrations, or designed for the individual as with self-paced instruction.

The determination of methods and materials for the learning activity can be as varied as the employer's imagination and available resources will allow. The employer may want to use charts, diagrams, manuals, slides, films, viewgraphs (overhead transparencies), videotapes, audiotapes, or simply blackboard and chalk, or any combination of these and other instructional aids. Whatever the method of instruction, the learning activities should be developed in such a way that the employees can clearly demonstrate that they have acquired the desired skills or knowledge.

Conducting the Training

With the completion of the steps outlined above, the employer is ready to begin conducting the training. To the extent possible, the training should be presented so that its organization and meaning are clear to the employees. To do so, employers or supervisors should:

(1) Provide overviews of the material to be learned.

(2) Relate, wherever possible, the new information or skills to the employee's goals, interests, or experience.

(3) Reinforce what the employees learned by summarizing the program's objectives and the key points of information covered.

These steps will assist employers in presenting the training in a clear, unambiguous manner.

In addition to organizing the content, employers must also develop the structure and format of the training. The content developed for the program, the nature of the workplace or other training site, and the resources available for training will help employers determine for themselves the frequency of training activities, the length of the sessions, the instructional techniques, and the individual(s) best qualified to present the information.

In order to be motivated to pay attention and learn the material that the employer or supervisor is presenting, employees must be convinced of the importance and relevance of the material. Ways to develop motivation include:

(1) Explaining the goals and objectives of instruction.

(2) Relating the training to the interests, skills, and experiences of the employees.

(3) Outlining the main points to be presented during the training session(s).

(4) Emphasizing the benefits of training (e.g., the employee will be better informed, more skilled, and thus more valuable both on the job and on the labor market, or the employee will, if he or she applies the skills and knowledge learned, be able to work at reduced risk).

An effective training program allows employees to participate in the training process and to practice their newly acquired skills or knowledge. This will help to ensure that they are learning the required knowledge or skills and permit correction if necessary. Employees can become involved in the training process by participating in discussions, asking questions, contributing their knowledge and expertise, and learning through hands-on experiences and role-playing exercises.

Evaluating Program Effectiveness

To make sure that the training program is accomplishing its goals, an evaluation of the training can be valuable. As one of its critical components, training should have a method of measuring the effectiveness of the training. The employer should develop a plan for evaluating the training session(s) when the course objectives and content are established. It should not be delayed until the training has been completed. Evaluation will help employers or supervisors determine the amount of learning achieved and whether an employee's performance has improved on the job. Among the methods of evaluating training are:

(1) **Student opinion.** Questionnaires or informal discussions with employees can help employers determine the relevance and appropriateness of the training program.

(2) **Supervisors' observations.** Supervisors are in good positions to observe an employee's performance both before and after the training and note improvements or changes.

(3) **Workplace improvements.** The ultimate success of a training program may be changes throughout the workplace that result in reduced injury or accident rates.

However it is conducted, an training evaluations can give employers the information necessary to decide whether or not the employees achieved the desired results, and whether the training session should be offered again at some future date.

Improving the Program

If after evaluation, it is clear that the training did not give the employees the level of knowledge and skill that was expected, it may be necessary to revise the training program or provide periodic retraining. At this point, asking questions of employees and of those who conducted the training may be of some help. Among the questions that could be asked are:

(1) Were parts of the content already known, and therefore, unnecessary?

(2) What material was confusing or distracting?

(3) Was anything missing from the program?

(4) What did the employees learn and what did they fail to learn?

It may be necessary to repeat steps in the training process, that is, to return to the first steps and retrace one's way through the training process. As the program is evaluated, the employer should ask:

(1) If a job analysis was conducted, was it accurate?

(2) Was any critical feature of the job overlooked?

(3) Were the important gaps in knowledge and skill included?

(4) Was material already known by the employees intentionally omitted?

(5) Were the instructional objectives presented clearly and concretely?

(6) Did the objectives state the level of acceptable performance that was expected of employees?

(7) Did the learning activity simulate the actual job?

(8) Was the learning activity appropriate for the kinds of knowledge and skills required on the job?

(9) When the training was presented, was the organization of the material and its meaning made clear?

(10) Were the employees motivated to learn?

(11) Were the employees allowed to participate actively in the training process?

(12) Was the employer's evaluation of the program thorough?

A critical examination of the steps in the training process will help employers to determine where course revision is necessary.

Matching Training to Employees

While all employees are entitled to know as much as possible about the safety and health hazards to which they are exposed, and employers should attempt to provide all relevant information and instruction to all employees, the resources for such an effort frequently are not, or are not believed to be, available. Thus, employers are often faced with the problem of deciding who is in the greatest need of information and instruction.

One way to differentiate between employees who have priority needs for training and those who do not is to identify employee populations that are at higher levels of risk. The nature of the work will indicate that such groups should receive priority for information on occupational safety and health risks.

Identifying Employees at Risk

One method of identifying employee populations at high levels of occupational risk (and thus in greater need of safety and health training) is to pinpoint hazardous occupations. Even within industries that are hazardous in general, there are some employees who operate at greater risk than others. In other cases, the hazardousness of an occupation is influenced by the conditions under which it is performed, such as noise, heat, or cold, or safety or health hazards in the surrounding area. In these situations, employees should be trained not only on how to perform their job safely, but also on how to operate within a hazardous environment.

A second method of identifying employee populations at high levels of risk is to examine the incidence of accidents and injuries, both within the company and within the industry. If employees in certain occupational categories are experiencing higher accident and injury rates than other employees, training may be one way to reduce that rate. In addition, thorough accident investigation can identify not only specific employees who could benefit from training, but also company-wide training needs.

Research has identified the following variables as being related to a disproportionate share of injuries and illnesses at the worksite on the part of employees:

(1) The age of the employee (younger employees have higher incidence rates).

(2) The length of time on the job (new employees have higher incidence rates).

(3) The size of the firm (in general terms, medium-size firms have higher incidence rates than smaller or larger firms).

(4) The type of work performed (incidence and severity rates vary significantly by Standard Industrial Classification (SIC) code).

(5) The use of hazardous substances (by SIC code).

These variables should be considered when identifying employee groups for training in occupational safety and health.

In summary, information is readily available to help employers identify which employees should receive safety and health information, education, and training, and who should receive it before others. Employers can request assistance in obtaining information by contacting such organizations as OSHA area offices, the Bureau of Labor Statistics, OSHA-approved state programs, state onsite consultation programs, the OSHA Office of Training and Education, or local safety councils.

Training Employees at Risk

Determining the content of training for employee populations at higher levels of risk is similar to determining what any employee needs to know, but more emphasis is placed on the requirements of the job and the possibility of injury. One useful tool for determining training content from job requirements is the Job Hazard Analysis described earlier. This procedure examines each step of a job, identifies existing or potential hazards, and determines the best way to perform the job in order to reduce or eliminate the hazards. Its key elements are:

(1) Job description

(2) Job location

(3) Key steps (preferably in the order in which they are performed)

(4) Tools, machines, and materials used

(5) Actual and potential safety and health hazards associated with these key job steps

(6) Safe and healthful practices, apparel, and equipment required for each job step

Material Safety Data Sheets (MSDS) can also provide information for training employees in the safe use of materials. These data sheets, developed by chemical manufacturers and importers, are supplied with manufacturing or construction materials and describe the ingredients of a product, its hazards, protective equipment to be used, safe handling procedures, and emergency first-aid responses. The information contained in these sheets can help employers identify employees in need of training (i.e., workers handling substances described in the sheets) and train employees in safe use of the substances. Material Safety Data Sheets are generally available from suppliers, manufacturers of the substance, or large employers who use the substance on a regular basis, or they can be developed by employers or trade associations. MSDSs are particularly useful for those employers who are developing training on chemical use as required by OSHA's Hazard Communication Standard.

Summary

In an attempt to assist employers with their occupational health and safety training activities, OSHA has developed a set of training guidelines in the form of a model. This model is designed to help employers develop instructional programs as part of their total education and training effort. The model addresses the questions of who should be trained, on what topics, and for what purposes. It also helps employers determine how effective the program has been and enables them to identify employees who are in greatest need of education and training. The model is general enough to be used in any area of occupational safety and health training, and allows employers to determine for themselves the content and format of training. Use of this model in training activities is just one of many ways that employers can comply with the OSHA standards that relate to training, and enhance the safety and health of their employees.

Chapter 5

Self-Inspections

Self-audits are an important component of an effective safety and health program and their purpose is to find and remedy hazardous conditions. Once the employer permanently identifies a hazardous condition, taking appropriate steps will prevent a recurrence of the violation or the occurrence of an accident or other losses. If a self-audit discloses a condition that is unsafe or is a violation of the OSH Act, presumably the employer would take action to correct the problem. In addition to human losses, the financial losses can be staggering–property damage, work stoppage, and worker's compensation among other things.

Inspections by a Competent Person

The standard requires that a competent person inspect, on a daily basis, excavations and the adjacent areas for possible cave-ins, failures of protective systems and equipment, hazardous atmospheres, or other hazardous conditions. If these conditions are encountered, exposed employees must be removed from the hazardous area until the necessary safety precautions have been taken. Inspections are also required after natural (e.g., heavy rains) or man-made events (e.g., blasting) that may increase the potential for hazards. The facts surrounding each excavation job must be evaluated to determine whether a competent person or qualified person is needed. While OSHA does not require that the inspections made by a competent person be documented, documentation is highly recommended, not just to prove compliance, but also as a means to prove that the daily inspections took place and create a safe and healthy worksite for all employees. The following statement is from the OSHA Technical Manual-Section V, Chapter 2: "Inspections shall be made by a competent person and *should* be documented." An example of a daily excavation condition log is located in Appendix A. Also, Appendix C (Site Assessment Questions) contains a list of questions that may be useful at an excavation.

Definition of a Competent Person

A competent person is defined by OSHA as "one who is capable of identifying existing and predictable hazards in the surroundings or working conditions which are unsanitary, hazardous, or dangerous to employees, and who has authorization to take prompt corrective measures to eliminate them" (OSHA 1926.32, Definitions). This definition is further clarified in the preamble for the rulemaking on excavations (54 Federal Register (FR) 45909) issued on October 31, 1989. That preamble states that what constitutes a "competent person" depends on the context in which the term is used. In order to be a "competent person" for the purposes of the excavation standard, one must have specific training in, and be knowledgeable about, soils analysis, the use of protective systems, and the requirements of the excavation standard. One who does not have this training or knowledge is not considered by OSHA to be capable of identifying existing and predictable hazards in excavation work, nor capable of taking the necessary corrective measures. By contrast, a "qualified" person or engineer, as defined in 1926.32(l), might have more technical expertise, but would not necessarily have expertise in hazard recognition or the authority to correct identified hazards.

Requirements for the Competent Person

The individual designated by management as the competent person for excavations must meet two qualifications:

(1) Be capable of identifying existing and predictable hazards.

(2) Be authorized to take prompt corrective measures to eliminate the hazards.

OSHA has defined how the qualifications to be a competent person can be met. In 1926.32(m),"qualified" is defined as "one who, by possession of a recognized degree, certificate, or professional standing, or who by extensive knowledge, training, and experience, has successfully demonstrated his ability to solve or resolve problems relating to the subject matter, the work, or the project." In 1926.32(i) a "designated person" is defined as a "person approved or assigned by the employer to perform a specific type of duty or duties or to be at a specific location or locations at the jobsite."

In August 1995, OSHA issued Standard Interpretations, a portion of which clarified the standard (1926.32(f)) that addresses competent persons at construction sites. The question was whether an intermediate employer or a subcontractor has the responsibility to employ competent persons at construction sites as called for in various standards, or whether it is OSHA's policy that a general contractor must also employ a competent person for each of the different trades or hazards at a construction site. In the interpretation, the question was only answered in general terms in view of the wide variety of fact patterns which multi-employer worksites present.

There are several specific standards, such as the crane standard and the excavation standard, that require the involvement of a competent person. Also, there is the general requirement found at 1926.20(b)(2)*:

* *1926.20(b)(2) - Such programs shall provide for frequent and regular inspections of the job sites, materials, and equipment to be made by competent persons designated by the employers.*

OSHA's response is that liability under the OSH Act may attach as a result of the employer's status as a creating, exposing, or controlling employer. It is also well settled that an employer's statutory obligations cannot be avoided through contractual arrangements. In this example, even if the contract (with the storage tank's owner) had provided for the instruction of Baker's employees, Baker could not contract away its legal duties to its employees or its ultimate responsibility under the act by requiring another party to perform them.

Another legal principle relevant to the inquiry was that the general contractor normally has the responsibility to assure that the other contractors fulfill their obligations with respect to employee safety, which affects the entire site. The general contractor is well situated to obtain abatement of hazards, either through its own resources or through its supervisory role with respect to other contractors. It is therefore reasonable to expect the general contractor to assure compliance with the standards insofar as all employees on the site are affected. Thus, OSHA will hold the general contractor responsible for violations it could reasonably have been expected to prevent or abate by reason of its supervisory capacity.

The Occupational Safety and Health Review Commission has also noted that the duty imposed on a general contractor is a reasonable one and that a general contractor will not be held liable for violations that it could not reasonably be expected to detect or prevent. The commission has also noted that, far from requiring the contracting employer to duplicate the safety efforts of the specialist, the act demands only that general contractors apprise themselves of which safety efforts their specialty subcontractors have chosen to make in completing their assignments.

How the Standards Apply to Experts, Specialists, and Consultants

When an employer contracts with a specialist, the employer is justified in relying on the specialist to protect against hazards related to the specialist's expertise, as long as the reliance is reasonable and the employer has no reason to foresee that the work will be performed unsafely. In other words, a general contractor may rely on a subcontractor if there is a sound basis for that reliance.

OSHA expects the employer to provide a competent person when an employer is engaged in an activity for which a competent person is required by a specific standard. In many situations, that employer will be a subcontractor rather than the general contractor. Other employers on the worksite, including the general contractor, who may have employees exposed to hazards associated with the activity need not provide a competent person of their own provided that they have taken appropriate steps to assure themselves that there is a competent person who is carrying out the prescribed duties. This point is illustrated not only by the decision discussed above but by another decision of the commission in a case when a citation was upheld against a formwork subcontractor whose employees worked in an excavation which it had not dug. The commission held that if the subcontractor did not employ a competent person of its own, it was obliged to assure itself that the requirements of the standard were being fulfilled by another party.

Shared Responsibility Between Contractors and Subcontractors

In addition to whatever responsibility it may have as an exposing employer, a general contractor would also share responsibility for those violations by its subcontractors that it could reasonably have detected and had corrected. OSHA would expect general contractors to take the steps necessary to satisfy themselves that subcontractors engaged in activities that call for the involvement of competent persons will fulfill that obligation as well as the general inspection obligation imposed by 1926.20(b)(2). Alternatively, a general contractor may decide to take responsibility for all or some portion of the competent person obligations that may arise at the worksite. To the extent that a general contractor undertakes, by contract or otherwise, to inspect for worksite hazards, the persons performing such duties obviously must have sufficient competence to be capable of detecting the hazardous and violative conditions that may arise at the particular worksite.

Good communications on a multi-employer worksite will prevent unnecessary duplication of effort while at the same time assure that each employer fulfills his obligations under the act.

Some important aspects of the letter of interpretation include the requirement for daily inspections of excavations, the adjacent areas, and protective systems by a competent person for:

a. Evidence of a situation that could result in possible cave-ins.

b. Indication of failure of protective systems or hazardous atmospheres.

c. Other hazardous conditions.

OSHA's Serious Violations: A Roadmap for Self-Inspections

The most frequent OSHA-cited serious excavation violations from January 1, 1990, to April 1, 1996, are shown in Table 5.1. The competent person should be cognizant of these violations and pay particular attention to each when performing daily inspections.

Table 5.1. Most Frequently Cited Serious Violations
January 1, 1990, to April 1, 1996

29 CFR 1926.652(a)(1)	(Protection in Excavations)
.651(k)(1)	(Inspections)
.651(j)(2)	(Loose Rock/Soil)
.651(c)(2)	(Means of Egress)
.651(d)	(Vehicular Traffic)
.651(k)(2)	(Inspections)
.651(h)(1)	(Water Accumulation)
.651(j)(1)	(Loose Rock/Soil)
.651(l)(2)*	(Walkways/Guardrails)
.651(e)	(Falling Loads)
.651(i)(3)	(Adjacent Structures)
.651(i)(3)	(Adjacent Structures)
.651(l)(1)*	(Walkways/Guardrails)
.652(b)	(Sloping/Benching Systems)
.651(i)(1)	(Adjacent Structures)
.652(c)	(Design/Protective Systems)
.652(g)(2)	(Shield Systems Requirements)
.652(g)(1)	(Shield Systems/General)
.651(b)(4)	(Underground Installations)
.651(g)(1)	(Hazardous Atmospheres)
.651(a)	(Surface Encumbrances)
.652(a)(2)	(Protective Systems)

* *29 CFR 1926.651(l)(2) was deleted by 59 FR 40730, dated August 9, 1994 (the publication of final rule Subpart M of Part 1926, Fall Protection).*

Excavation requirements for fall protection are stated in 29 CFR 1926.501 (Duty to Have Fall Protection) and are as follows:

*"**Excavations.***

1926.501(b)(7)(i)

Each employee at the edge of an excavation 6 feet (1.8 m) or more in depth shall be protected from falling by guardrail systems, fences, or barricades when the excavations are not readily seen because of plant growth or other visual barrier;

1926.501(b)(7)(ii)

Each employee at the edge of a well, pit, shaft, and similar excavation 6 feet (1.8 m) or more in depth shall be protected from falling by guardrail systems, fences, barricades, or covers."

Effective control measures should be used for the frequently cited serious excavation violations discussed above. Following are just two examples of such controls:

(1) The competent person could develop a checklist enumerating the items listed and use the list to identify and correct unsafe or unhealthy conditions that exist on a particular worksite.

(2) All excavations including trenches that are 5 feet (1.52 m) in depth or greater (1926.652) must be shored or sloped. For excavations less than 5 feet (1.52 m) in depth, the competent person is required to examine the excavation for potential cave-in hazards and determine if protection is needed.

Competent persons have an important role and trenching and excavation work is dependent on these specialized employees because its highly technical nature as well as its inherent hazards require a greater level of training and experience than a normal worker would possess. The items below specify the trenching and excavation activities where a competent person is necessary.

Aggregated data for other most frequently cited standards related to excavation work from January 1,1990, to April 1, 1996, are presented in Table 5.2.

Inspect and Evaluate Protective Systems and Equipment

This standard provides requirements that allow employers flexibility in developing programs that provide effective protection for employees working in excavations. Daily inspections and evaluations of excavations, the adjacent areas, and protective systems must be made by a competent person for evidence of a situation that could result in cave-ins, failure of protective systems, hazardous atmospheres, or other hazardous conditions. The following are examples of inspection guidelines that can be used to ensure compliance in accordance with 29 CFR 1926, Subpart P, Excavations.

Monitoring water removal equipment and operations (1926.651(h)(2)). "If water is controlled or prevented from accumulating by the use of water removal equipment, the water removal equipment and operations shall be monitored by a competent person to ensure proper operation."

Inspecting excavations subject to runoff from heavy rains to determine need for diversion ditches, dikes, or other suitable protection (1926.651(h)(3)). "If excavation work interrupts the natural drainage of surface water (such as streams), diversion ditches, dikes, or other suitable means shall be used to prevent surface water from

Table 5.2. Most Frequently Cited Standards Related to Excavation Work

Standard	Title	Total Violations	Serious	Willful	Repeat	Uncl	Other
1926.21b	The employer shall instruct each employee in the recognition and avoidance of unsafe conditions	7975	6706	50	157	7	1055
1926.100a	Head protection, where there is a possible danger of head injury.	6912	6012	9	233	3	655
1926.20b1	Employer responsibility to initiate and maintain safety and health programs.	5345	3525	16	100	2	1702
1926.20b2	Employer responsibility to provide for frequent and regular inspections by designated competent persons.	4385	3721	17	86	2	559
1926.1052c1	Stair rail and handrail along each unprotected edge.	3614	2348	4	14	1	1157
1926.50	First aid care shall be available in the absence of an infirmary or other location that is reasonably accessible, first aid supplies shall be accessible and telephone numbers posted.	3358	834	0	34	2	2488
1926.28a	The employer is responsible for the wearing of appropriate personal protective equipment.	3240	2703	28	71	3	435
1926.102a1	Eye and face protection shall be provided.	2131	1813	3	28	1	286
1926.405a2ii e&f	Temporary lights shall be protected from breakage, not suspended by their cords and extension cord.	1815	983	0	36	0	796
1926.95a	Personal protective equipment shall be provided, used and maintained in a sanitary and reliable condition.	921	765	24	15	7	110

entering the excavation and to provide adequate drainage of the area adjacent to the excavation. Excavations subject to runoff from heavy rains will require an inspection by a competent person and compliance with paragraphs (h)(1) and (h)(2) of this section."

Determining cave-in potential to assess need for shoring or other protective system (1926.652(a)(1)). "Each employee in an excavation shall be protected from cave-ins by an adequate protective system designed in accordance with paragraph (b) or (c) of this section except when: 1926.652(a)(1)(i) Excavations are made entirely in stable rock; or 926.652(a)(1)(ii) Excavations are less than 5 feet (1.52 m) in depth and examination of the ground by a competent person provides no indication of a potential cave-in."

Examining damaged material or equipment used for protective systems to determine its suitability for continued use (1926.652(d)(3)). "When material or equipment that is used for protective systems is damaged, a competent person shall examine the material or equipment and evaluate its suitability for continued use. If the competent person cannot assure the material or equipment is able to support the intended loads or is otherwise suitable for safe use, then such material or equipment shall be removed from service, and shall be evaluated and approved by a registered professional engineer before being returned to service." Shields or other protection systems, damaged to the extent that the structural integrity or the ability of the system to perform as designed is affected, cannot be used until repaired and recertified by a registered professional engineer or the manufacturer of the system. Minor repairs or normal maintenance done in accordance with manufacturer's specifications do not need recertification.

Classifying soil and rock deposits, by both visual analysis and by testing, to determine appropriate protection; re-classifying, if necessary, based on changing conditions (1926 Subpart P App A). "(a) Scope and application - (1) Scope. This appendix describes a method of classifying soil and rock deposits based on site and environmental conditions, and on the structure and composition of the earth deposits. The appendix contains definitions, sets forth requirements, and describes acceptable visual and manual tests for use in classifying soils."

Determining the appropriate slope of an excavation to prevent collapse due to surcharge loads from stored material or equipment, operating equipment, adjacent structures, or traffic, and assuring that such slope is achieved (1926 Subpart P App B(c)(3)(iii)). "When surcharge loads from stored material or equipment, operating equipment, or traffic are present, a competent person shall determine the degree to which the actual slope must be reduced below the maximum allowable slope, and shall assure that such reduction is achieved. Surcharge loads from adjacent structures shall be evaluated in accordance with 1926.651(i) Stability of Adjacent Structures."

"1926.651(i) Stability of Adjacent Structures.

1926.651(i)(1)

Where the stability of adjoining buildings, walls, or other structures is endangered by excavation operations, support systems such as shoring, bracing, or underpinning shall be provided to ensure the stability of such structures for the protection of employees."

Registered professional engineer approval is not required when the excavation is not "reasonably expected to pose a hazard to employees." In situations where it is reasonably expected to pose a hazard, approval by a registered professional engineer is not required when a support system, such as underpinning, is provided to ensure the safety of employees and the stability of the structure, or the excavation is in stable rock.

"1926.651(i)(2)

Excavation below the level of the base or footing of any foundation or retaining wall that could be reasonably expected to pose a hazard to employees shall not be permitted except when:

1926.651(i)(2)(i)

A support system, such as underpinning, is provided to ensure the safety of employees and the stability of the structure; or

1926.651(i)(2)(ii)

The excavation is in stable rock; or

1926.651(i)(2)(iii)

A registered professional engineer has approved the determination that the structure is sufficiently removed from the excavation so as to be unaffected by the excavation activity; or

1926.651(i)(2)(iv)

A registered professional engineer has approved the determination that such excavation work will not pose a hazard to employees.

1926.651(i)(3)

Sidewalks, pavements and appurtenant structure shall not be undermined unless a support system or another method of protection is provided to protect employees from the possible collapse of such structures."

Inspect and Evaluate Trench and Protective Systems

Authorizing immediate removal of employees from the hazardous area where evidence of possible cave-in, failure of protective systems, hazardous atmospheres, or other hazardous conditions exists (1926.651(k)(2)). "Where the competent person finds evidence of a situation that could result in a possible cave-in, indications of failure of protective systems, hazardous atmospheres, or other hazardous conditions, exposed employees shall be removed from the hazardous area until the necessary precautions have been taken to ensure their safety."

Chapter 6

Utility Hazards

The unexpected discovery of an unmarked, undetected underground utility line is one of the more dangerous aspects of the construction industry. These "phantom" pipes, wires, or conduits are frequently missing from as-built or other record drawings and often elude detection via "traditional craft" surface geophysical methods, such as pipe and cable locators and ground penetrating radar. Underground lines are often energized, pressurized, or contain radioactive or other hazardous substances. If they are accidentally damaged, they can inflict serious injury on workers in the vicinity of the site. Utility service to facilities in the general area can also be disrupted for significant lengths of time. Facility and nuclear safety may also be adversely affected because of power loss to safety-related systems.

These incidents often involve various types of equipment (hand tools, jackhammers, backhoes) and the full range of utility lines and pipes (natural gas, electric, water, industrial waste). It is also probable that substantial additional costs and project delays are incurred as a result of last minute construction discoveries of unknown or misrepresented underground utilities.

Improper planning and carelessness cause the majority of these incidents. In many cases, workers do not maintain adequate safety margins during excavation activities and do not fully understand excavation procedures and potential hazards. Finally, people involved in these events often over rely on facility as-built or other record drawings and do not properly use or interpret the data from surface geophysical equipment used for utility detection and location of other hidden hazards.

The following incidents are adapted from actual cases.

Worker Seriously Injured when Jackhammer Hits Underground 13,200-Volt Line

A mason tender was severely injured when he hit a buried 13,200-volt electrical power line with a jackhammer. The worker received serious burns, was rendered unconscious, and at this writing is still in a coma. Electrical power was disrupted in the facility and in the surrounding buildings. The excavation project required workers to cut a 3-foot deep hole in the basement of a building for a sump pump project. A foreman and mason tender were assigned to the project. They alternated use of a shovel with the use of a jackhammer to remove soil and tuff rock. Both wore standard protective clothing, including leather gloves, safety shoes, hard hats, and eye and ear protection. However, their protective clothing was not designed for high-voltage work.

After the event, facility managers determined that safety engineers had not performed a review to identify health and safety hazards before work began. The workers were not aware that an excavation permit and an environment, safety, and health review were required for concrete cutting and sub-slab excavation inside a building.

Investigators identified the following contributing factors in the report.

- Tighter project controls and engineering reviews could have identified the hazards before the start of the work.
- The support organization assigned to manage the project was ill-equipped to perform this type of complex facility modification work.
- The use of a standing work order to guide the work did not provide an adequate description of the facility or the work hazards. Standing work orders are neither designed nor intended to be used for large maintenance tasks, complex facility modifications, or major construction activities.
- Schedule pressures and commitments allowed the work to proceed without sufficient detail and supervision.
- Misinterpretation of local excavation permit procedures resulted in requests for permits only for areas located outside buildings. Because of this, appropriate measures were not taken to determine the location of dangerous underground utilities and prevent exposure of employees to potential hazards inside buildings.
- The use of an electromagnetic detection device would have located the buried electrical conduit. This was shown during the accident investigation by using a detection device at the scene of the accident.

Using personal protective equipment designed for high voltage work also could have minimized the injuries. The accident victim was not wearing the proper gloves at the time of the accident, and there was no evidence that the victim had ever been trained or certified in the use and care of personal protective equipment.

A review of the original construction plans and specifications indicated that the conduit encasing the electrical cable should have been made of rigid steel. The actual installation was an asphalt-impregnated, fiber-based conduit. A steel conduit would have provided another barrier of protection against contact with the energized cable.

Finally, the compressive strength of the concrete surrounding the conduit was relatively low. Because of this low compressive strength, the difference between the tuff rock and the concrete could easily have been overlooked by the workers, allowing them to penetrate the concrete with the jackhammer without realizing they had done so.

Natural Gas Pipe Hit by Backhoe, Site Area Evacuated

A construction contractor hit and ruptured a 3-inch polyvinyl chloride (PVC) natural gas pipe. Three construction employees were using both hand and backhoe excavating methods when they hit the natural gas pipe approximately 20 inches below grade. The backhoe cut into the pipe, causing a discharge of natural gas. The immediate area and one adjacent building were evacuated. The natural gas release created the potential for a serious explosion.

Planners provided composite drawings to the contractor along with the digging permit. These utility drawings included a composite of all known buried utilities within the proposed excavation site. Because of the number of utilities in the area, the drawings were complex and cluttered, making them very difficult to interpret. The contractor did not note the location of the natural gas pipe on the utility drawing; therefore, the location was not verified before excavation began.

Electromagnetic detection was not used before beginning excavation. Using it could have prevented the incident because an electrically conductive tracer wire was properly installed above the pipe. Finally, the gas pipe was not properly marked with caution tape as the final warning barrier.

Corrective actions included the following.

- Managers modified the digging permit process to more closely reflect the requirements of the local public utility company. Modifications included incorporating a new requirement to physically mark the surface of the excavation area to designate buried utilities. Additional new requirements included conducting onsite meetings to review the utilities in the area, use of hand-digging, and other appropriate requirements.
- Managers initiated steps to ensure that only personnel trained in the proper techniques and equipment will be used to locate underground utilities. They also ensured that the proper detection equipment was obtained.

- Managers initiated a study to determine the feasibility of adding penalty clauses to construction contracts and pursuing actual damages to recover all costs associated with damaged utility lines. They also directed examining existing contractor evaluation clauses to limit an offending contractor's ability to bid on future work.

Water Line Damaged While Excavating

A site power crew accidentally cut into an underground 6-inch PVC water line during installation of guy anchors for power line poles. A ground-penetrating radar (GRP) survey had previously identified all the underground utilities, including the water line, and marked them with red flags and paint marks on the ground. The power crew stopped work and notified the facility manager, who further informed the Site Utilities Division about the water line breach. The resulting loss of pressure in the water line caused disruption to a number of site operations. Improper construction practices can seriously impact safety-related critical systems and endanger personnel safety and health.

The workers used a power auger instead of the required hand-digging in the vicinity of the marked utilities and inadvertently ruptured the water line. Investigators determined that water pressure was lost in the three other lines. The low water pressure interfered with normal operations, including cooling water to air compressors and chillers. One facility manager ordered the process air primary and backup compressors shut down to prevent damage resulting from dead ending on closed dampers. The facility manager also stopped breathing air compressors and placed the facility in a limiting condition of operations (LCO). High Volume Air Activity Monitor (HVAAM) exhausters were also shut down, requiring implementation of respiratory protection procedures.

Investigators determined that the responsible construction discipline engineer performed a system walk-down with a sketch that indicated the layout of the survey interferences. Investigators discovered that this field map was not included in the work package. Investigators also determined that the work clearance permit of the work package identified "Excavation" work, which provides a reference to instructions for digging around the marked utilities, but the workers overlooked that requirement. Investigators further found that the work crew did not perform hand-digging to expose all underground utilities that needed protection in the construction area. Investigators identified the following breakdowns of the Integrated Safety Management system.

The potential loss of domestic water and its impact on facility operations were not adequately considered nor understood during work planning. Controls to mitigate normal hazards associated with excavations, as specified in the site construction safety manual, were not followed. No steps were taken to verify compliance with site-specific requirements or to include additional controls, based on the potential for an adverse impact on multiple operations. The facility management held a meeting and identified the following corrective actions.

- Issue a work package to repair and flush the broken water line and verify water quality after restoration of normal service.
- Hand dig to verify actual location of underground lines shown on drawings and site maps and locate or verify the absence of any interference by using ground penetrating radar.
- Review the current protocol for contacting shift operations managers during evolutions or events impacting or with a potential to impact facility operations, emphasizing the inclusion of all affected divisions and facilities.
- Ensure that the site's excavation procedures are required reading for all personnel tasked with digging operations.

Underground Waste Transfer Pipe Incorrectly Described as an Electrical Conduit

A worker at a nuclear plant uncovered an abandoned underground radioactive waste transfer pipe. The work package incorrectly described the pipe as an electrical conduit. The worker, who was hand-digging a trench at the time, received an estimated dose of 5 to 7 millirem (mrem) before the pipe was identified as radioactive and posted.

While preparing the work package, a planner mistakenly read the drawing as showing an underground electrical conduit instead of an underground waste transfer pipe. Because of this error, planners did not include the requirement for radiological controls with the excavation permit. Fortunately, a Health Physics (HP) supervisor decided to include continuous health physics monitoring because of past history with contaminated soil in the area. He assigned a Health Physics Technician (HPT) to the site. While the HPT was at lunch, the worker continued to dig, in violation of the excavation permit, and uncovered the waste pipe. Upon his return, the HPT determined that the radiation level was 40 mrem at contact with the pipe. The HPT immediately stopped the work and secured the area.

The waste transfer pipe was shown on the project drawings and had previously been located using ground penetrating radar. However, because the excavators were mistakenly expecting to find an electrical conduit rather than the waste transfer pipe, previous location of the line did not help prevent the event.

Corrective actions included the following.

- Management tasked onsite construction managers with reviewing and segregating their excavation work into separate permits by location or system.

- Management determined that inadequate posting for radioactive and contaminated areas (including buried radioactive pipelines) was a site-wide issue, and changes to postings are being implemented.
- Management distributed lessons learned conveying the meaning of, and requirements for, continuous HPT coverage to construction personnel and managers.

Management made significant improvements in the site excavation program. They developed a new site-wide excavation procedure and established an excavation coordinator as the single point-of-contact for the excavation process. In an article entitled "The Engineer's View" from the November/December 1995 issue of Underground Focus, James H. Anspach, Senior Geophysicist at So-Deep, Inc., describes the program as follows:

"... the Excavation Coordinator (EC) now functions much as a one-call center, although with many additional duties. Not only does the EC take calls for construction, but also for design. The EC makes certain:

- Appropriate utility records are investigated and reviewed.
- Appropriate surface geophysical methods are employed to identify active, abandoned, metallic, non-metallic, recorded, and non recorded utilities.
- Utilities are depicted on design plans.
- Utilities are marked on the ground surface prior to construction.
- Utilities discovered during construction that slipped through the surface geophysics and records review are investigated.
- As-built drawings are revised to reflect physical configuration.
- Changes or additions to the utility system during construction are as-built and retained in a retrievable database."

It is important at similar sites for all contractors involved in excavation to participate in a centrally controlled excavation process. If they fail to do so, the effectiveness of preventing underground utility damage can be seriously jeopardized.

Electrical Conduit Punctured During Concrete Floor Excavation

A shift technical engineer discovered a punctured electrical conduit at a river site. Construction workers damaged the conduit when they removed a section of a concrete floor with a jackhammer in preparation for hand excavation of a trench. The workers were not aware that they had damaged the conduit.

Construction personnel used ground-penetrating radar to locate buried pipes and conduits before the excavation was started. They detected an interference (metallic object) 16 inches below the floor slab. Trenching procedures dictated that a 1-foot clearance zone be established around each interference to identify areas that must be hand excavated, and from which power equipment must be excluded, to minimize the risk of personnel injury and damage to underground services during excavation. However, excavators misinterpreted the requirements' for interference point clearance zones. They were also unaware that the 1 - foot requirement applied to the vertical as well as the horizontal direction. This meant that the conduit could have been as shallow as 4 inches or as deep as 28 inches.

Facility investigators later determined that the conduit was 5 inches below the top of the slab. The planners and workers did not recognize that the 1-foot clearance requirement was applicable in this case, and did not consider it necessary to positively identify the interference or take measures to ensure that the worksite was safe. This occurred because they thought the interference was located with a high degree of accuracy at a depth of 16 inches, and because they only planned to cut and remove a section of the concrete slab that they thought was well above the interference.

Corrective actions included the following.

- Management reinforced lockout/tagout requirements for digging operations.
- Management directed more thorough engineering reviews of work control packages to identify buried conduits and pipes.
- Management revised trenching procedures to clearly state the clearance zone requirements when using ground-penetrating radar.
- Trainers instructed excavators on the limitations of ground-penetrating radar.

Costs of Excavation Events

The economic cost associated with excavation events is significant. An article in *Underground Focus* magazine states that "when a damage to a utility occurs during construction, the costs of investigating that damage... average about $50,000. The damaged item could be just a 4-foot piece of scrap wire or an abandoned clay drainage pipe. All that matters is that something has been damaged. All activity ceases until the investigation to identify the damaged item is complete."

Three primary contributors to excavation-related events were identified:

- Inability to detect underground utilities because of over reliance on as-built drawings.
- Schedule pressures leading to lack of use of hand-digging where appropriate.
- Inability to detect underground utilities caused primarily by ineffective or lack of use of electromagnetic and other detection devices.

Recommendations

The following recommendations are based on the above analysis.

A comprehensive underground utility detection and marking program should be implemented at each facility. Program developers should strongly consider creating a central excavation coordinator who would be a single point-of-contact for the excavation process. In addition, the program should ensure that data supplied by subsurface utility engineers is utilized during construction projects. Procedures should be developed, validated, and used in all situations involving excavation activities.

One of the primary objectives of an underground utility detection program should be to identify, implement, and maintain barriers that effectively prevent personnel injury, equipment damage, and degradation of nuclear and other safety margins during excavation activities.

State-of-the-art detection equipment should be obtained and operators should be trained in its proper use. The equipment must be properly calibrated, and equipment limitations must be fully understood. Excavation equipment should be used properly and cautiously, and it should never be used in place of hand-digging in the vicinity of identified utility lines.

As-built drawings should be kept up to date to the extent possible. Because of the lack of configuration control at many facilities, as-built drawings should not be relied upon as the only source for the accurate location of underground utility systems. Local public utility companies should be contacted and information should be shared. Many localities have comprehensive programs in place that require calling a single point of contact for any utility excavation, such as "One Call" and "Miss Utility." Such programs can be used as models for site/local programs.

Successes at other facilities should be publicized and made available throughout the complex. Managers should identify and review lessons learned and apply them where applicable.

Chapter 7

Guide to OSHA Compliance and Hazard Control

The Occupational Safety and Health Administration issued its first Excavation and Trenching Standard in 1971 to protect workers from excavation hazards. Since then, OSHA has amended the standard several times to increase worker protection and to reduce the frequency and severity of excavation accidents and injuries. To better assist excavation firms and contractors, OSHA completely updated the existing standard to simplify many of the existing provisions, add and clarify definitions, eliminate duplicate provisions and ambiguous language, and give employers added flexibility in providing protection for employees. The revised standard provides several useful appendices. One appendix describes a consistent method of soil classification. Others offer sloping and benching requirements, pictorial examples of shoring and shielding devices, timber tables, hydraulic shoring tables, and selection charts that provide a graphic summary of the requirements contained in the standard. This chapter highlights the requirements in the updated standard for excavation and trenching operations, provides methods for protecting employees against cave-ins, and describes safe work practices for employees.

This chapter contains the following three OSHA excavation standards along with Appendices A through F of 29 CFR 1926 Subpart P.

1. – 29 CFR 1926.650 – Scope, Application, and Definitions Applicable to this Subpart.
2. – 29 CFR 1926.651 – Specific Excavation Requirements
3. – 29 CFR 1926.652 – Requirements for Protective Systems

a. – 29 CFR 1926 Subpart P Appendix A – Soil Classification
b. – 29 CFR 1926 Subpart P Appendix B – Sloping and Benching
c. – 29 CFR 1926 Subpart P Appendix C – Timber Shoring for Trenches
d. – 29 CFR 1926 Subpart P Appendix D – Aluminum Hydraulic Shoring for Trenches
e. – 29 CFR 1926 Subpart P Appendix E – Alternatives to Timber Shoring
f. – 29 CFR 1926 Subpart P Appendix F – Selection of Protective Systems

Also included in this chapter are the author's comments (italicized text) on OSHA interpretations for the standards and appendices. This informational discussion is intended by the author to provide a generic, non-exhaustive overview of the excavation standards. This discussion outlines responsibilities associated with the excavation standards set forth in the OSHA standards and interpretations themselves, and in the *Occupational Safety and Health Act of 1970.* Moreover, because interpretations and enforcement policy may change over time, the reader should consult future administrative interpretations and decisions by the Occupational Safety and Health Review Commission and the courts for additional guidance on OSHA compliance requirements. However, OSHA's enforcement policy with regard to confined spaces at construction sites has not changed with the promulgation of the general industry regulation. In those instances where a hazard is addressed by an existing part 1926 standard, OSHA will continue to cite the specific standard. In those cases where a hazard is observed that is not addressed by an existing specific construction standard but is addressed in the American National Standards Institute's Z117.1 consensus standard, OSHA will continue to cite under 5(a)(1) of the act provided the conditions for citing the general duty clause are present. However, in no circumstance would OSHA cite a construction contractor under 1926.21 and 1926.651 for failure to comply with the requirements of a new general industry rule.

4.A. – 29 CFR 1926.650 - Scope, Application, and Definitions Applicable to this Subpart.

- Part Number: 1926
- Part Title: Safety and Health Regulations for Construction
- Subpart : P
- Subpart Title: Excavations
- Standard Number: 1926.650
- Title: Scope, Application, and Definitions Applicable to this Subpart.

1926.650(a)

Scope and Application. This subpart applies to all open excavations made in the earth's surface. Excavations are defined to include trenches.

1926.650(b)

"Accepted engineering practices" means those requirements which are compatible with standards of practice required by a registered professional engineer.

"Aluminum Hydraulic Shoring" means a pre-engineered shoring system comprised of aluminum hydraulic cylinders (crossbraces) used in conjunction with vertical rails (uprights) or horizontal rails (wales). Such system is designed specifically to support the sidewalls of an excavation and prevent cave-ins.

"Bell-bottom pier hole" means a type of shaft or footing excavation, the bottom of which is made larger than the cross section above to form a belled shape.

"Benching (Benching system)" means a method of protecting employees from cave-ins by excavating the sides of an excavation to form one or a series of horizontal levels or steps, usually with vertical or near-vertical surfaces between levels.

"Cave-in" means the separation of a mass of soil or rock material from the side of an excavation, or the loss of soil from under a trench shield or support system, and its sudden movement into the excavation, either by falling or sliding, in sufficient quantity so that it could entrap, bury, or otherwise injure and immobilize a person.

"Competent person" means one who is capable of identifying existing and predictable hazards in the surroundings, or working conditions which are unsanitary, hazardous, or dangerous to employees, and who has authorization to take prompt corrective measures to eliminate them.

Excavation standards refer to duties of a competent person in §§1926.651(c)(1)(i), (h)(2) and (3), and (k)(1) and (2), and 1926.652(a)(1)(ii) and (d)(3).

"Cross braces" mean the horizontal members of a shoring system installed perpendicular to the sides of the excavation, the ends of which bear against either uprights or wales.

"Excavation" means any man-made cut, cavity, trench, or depression in an earth surface, formed by earth removal.

"Faces" or "sides" means the vertical or inclined earth surfaces formed as a result of excavation work.

"Failure" means the breakage, displacement, or permanent deformation of a structural member or connection so as to reduce its structural integrity and its supportive capabilities.

"Hazardous atmosphere" means an atmosphere which by reason of being explosive, flammable, poisonous, corrosive, oxidizing, irritating, oxygen deficient, toxic, or otherwise harmful, may cause death, illness, or injury.

"Kickout" means the accidental release or failure of a cross brace.

"Protective system" means a method of protecting employees from cave-ins, from material that could fall or roll from an excavation face or into an excavation, or from the collapse of adjacent structures. Protective systems include support systems, sloping and benching systems, shield systems, and other systems that provide the necessary protection.

"Ramp" means an inclined walking or working surface that is used to gain access to one point from another, and is constructed from earth or from structural materials such as steel or wood.

"Registered Professional Engineer" means a person who is registered as a professional engineer in the state where the work is to be performed. However, a professional engineer, registered in any state is deemed to be a "registered professional engineer" within the meaning of this standard when approving designs for "manufactured protective systems" or "tabulated data" to be used in interstate commerce.

"Sheeting" means the members of a shoring system that retain the earth in position and in turn are supported by other members of the shoring system.

"Shield (Shield system)" means a structure that is able to withstand the forces imposed on it by a cave-in and thereby protect employees within the structure. Shields can be permanent structures or can be designed to be portable and moved along as work progresses. Additionally, shields can be either premanufactured or job-built in accordance with 1926.652(c)(3) or (c)(4). Shields used in trenches are usually referred to as "trench boxes" or "trench shields."

"Shoring (Shoring system)" means a structure such as a metal hydraulic, mechanical or timber shoring system that supports the sides of an excavation and which is designed to prevent cave-ins.

"Sides." See "Faces."

"Sloping (Sloping system)" means a method of protecting employees from cave-ins by excavating to form sides of an excavation that are inclined away from the excavation so as to prevent cave-ins. The angle of incline required to prevent a cave-in varies with differences in such factors as the soil type, environmental conditions of exposure, and application of surcharge loads.

"Stable rock" means natural solid mineral material that can be excavated with vertical sides and will remain intact while exposed. Unstable rock is considered to be stable when the rock material on the side or sides of the excavation is secured against caving-in or movement by rock bolts or by another protective system that has been designed by a registered professional engineer.

"Structural ramp" means a ramp built of steel or wood, usually used for vehicle access. Ramps made of soil or rocks are not considered structural ramps.

"Support system" means a structure such as underpinning, bracing, or shoring, which provides support to an adjacent structure, underground installation, or the sides of an excavation.

"Tabulated data" means tables and charts approved by a registered professional engineer and used to design and construct a protective system.

"Trench (Trench excavation)" means a narrow excavation (in relation to its length) made below the surface of the ground. In general, the depth is greater than the width, but the width of a trench (measured at the bottom) is not greater than 15 feet (4.6 m). If forms or other structures are installed or constructed in an excavation so as to reduce the dimension measured from the forms or structure to the side of the excavation to 15 feet (4.6 m) or less (measured at the bottom of the excavation), the excavation is also considered to be a trench.

"Trench box." See "Shield.

"Trench shield." See "Shield."

"Uprights" means the vertical members of a trench shoring system placed in contact with the earth and usually positioned so that individual members do not contact each other. Uprights placed so that individual members are closely spaced, in contact with or interconnected to each other, are often called "sheeting."

"Wales" means horizontal members of a shoring system placed parallel to the excavation face whose sides bear against the vertical members of the shoring system or earth.

4.B. – 29 CFR 1926.651 - Specific Excavation Requirements

- Part Number: 1926
- Part Title: Safety and Health Regulations for Construction
- Subpart : P
- Subpart Title: Excavations
- Standard Number: 1926.651
- Title: Specific Excavation Requirements

(a) Surface Encumbrances

All surface encumbrances that are located so as to create a hazard to employees shall be removed or supported, as necessary, to safeguard employees.

To meet this requirement, employers should identify (in writing) the location of the excavation and the specific operations and/or conditions. Any observed hazards should be described along with any controls developed.

(b) Underground Installations

The estimated location of utility installations, such as sewer, telephone, fuel, electric, water lines, or any other underground installations that reasonably may be expected to be encountered during excavation work, shall be determined prior to opening an excavation.

It is imperative to contact the local utility company prior to any digging. Careless digging can cause disruption of utility service or property damage, but more importantly, it can result in serious injury or death. Be sure to allow enough time for the project to request the location of utilities and or cable and phone lines. In accordance with the Department of Transportation Regulation 40 CFR 192.614, Damage Prevention Procedure Program, a call should be made to the county or parish one-call system in which the excavation is being made.

Utility companies or owners shall be contacted within established or customary local response times, advised of the proposed work, and asked to establish the location of the utility underground installations prior to the start of actual excavation. When utility companies or owners cannot respond to a request to locate underground utility installations within 24 hours (unless a longer period is required by state or local law), or cannot establish the exact location of these installations, the employer may proceed, provided the employer does so with caution, and provided detection equipment or other acceptable means to locate utility installations are used.

Utility companies or owners must be advised of any proposed work and asked to establish the location of the utility underground installations prior to the start of an actual excavation. This notification must be made within established or customary local response times. Once notification has been made, it is important to document all communications and not start work until the installation locations are marked.

When excavation operations approach the estimated location of underground installations, the exact location of the installations shall be determined by safe and acceptable means.

Maintain awareness of the installation location marks while digging and avoid machine digging within 2 feet of marked areas unless the utilities have been identified and adequately exposed and protected.

While the excavation is open, underground installations shall be protected, supported or removed as necessary to safeguard employees.

Identify any specific operations with underground installations requiring controls to be developed to protect employees from identified hazards. Document all hazards identified and the controls that were implemented.

(c) Access and Egress

Structural ramps.

Structural ramps that are used solely by employees as a means of access or egress from excavations shall be designed by a competent person. Structural ramps used for access or egress of equipment shall be designed by a competent person qualified in structural design, and shall be constructed in accordance with the design.

When structural ramps are used, either for means of access or egress for employee or equipment use, documentation of the design and construction approval by a competent person (qualified in structural design if ramps are used for access or egress of equipment) must be maintained on or near the worksite. The documentation should include the location, specific operations and/or conditions, and any identified hazards.

"Qualified" means one who, by possession of a recognized degree, certificate, or professional standing, or who by extensive knowledge, training, and experience, has successfully demonstrated his or her ability to solve or resolve problems relating to the subject matter, the work, or the project.

A qualified person must have a recognized degree, certificate, etc., or extensive experience and ability to solve the subject problems at the worksite. This is why .651(f) requires that supporting systems shall be designed by a qualified person. There may be a requirement for more technical or engineering knowledge here.

Ramps and runways constructed of two or more structural members shall have the structural members connected together to prevent displacement.

When ramps and runways have two or more members, documentation by a competent person is required to show how the structural members are connected together to prevent displacement.

Structural members used for ramps and runways shall be of uniform thickness.

If an incident occurs involving structural members of ramps and runways, part of the investigation will be to determine if the members were of uniform thickness.

Cleats or other appropriate means used to connect runway structural members shall be attached to the bottom of the runway or shall be attached in a manner to prevent tripping.

A competent person must approve in the original documentation the appropriate means of connecting runway structural members.

Structural ramps used in lieu of steps shall be provided with cleats or other surface treatments on the top surface to prevent slipping.

The location, specific operations and/or conditions, and any identified hazards should be documented. An example of such a hazard is when structural ramps used in lieu of steps are not provided with cleats or other surface treatments on the top surface to prevent slipping.

Means of egress from trench excavations. A stairway, ladder, ramp or other safe means of egress shall be located in trench excavations that are 4 feet (1.22 m) or more in depth so as to require no more than 25 feet (7.62 m) of lateral travel for employees.

Failure to ensure that adequate egress such as a ladder is used to enter and exit a trench is a willful violation. Measurements should be made as necessary and the means of egress (if any) provided should be documented. A ladder, stairway, ramp, or other safe means of access is required only when the trench is 4 feet or more in depth. Since this section addresses means of access and egress, the more general requirement in the stairways and ladders subpart is not applicable. Paragraphs 1926.1053(b)(6) and (7) address ladder footing displacement, which is not normally a problem in trenches. However, if a ladder needs to be secured against tipping in an excavation, it may be secured to a shield or member of a protective structure provided the ladder does not alter the effectiveness of the protective system.

(d) Exposure to Vehicular Traffic

Employees exposed to public vehicular traffic shall be provided with, and shall wear, warning vests or other suitable garments marked with or made of reflectorized or high-visibility material.

The location, specific operations and/or conditions, and any identified hazards and controls should be documented. A description of what (if anything) employees are required to wear when exposed to public vehicular traffic should also be documented.

Figure 6.1. Vehicular Traffic

(e) Exposure to falling loads

No employee shall be permitted underneath loads handled by lifting or digging equipment. Employees shall be required to stand away from any vehicle being loaded or unloaded to avoid being struck by any spillage or falling materials. Operators may remain in the cabs of vehicles being loaded or unloaded when the vehicles are equipped, in accordance with 1926.601(b)(6), to provide adequate protection for the operator during loading and unloading operations.

The equipment being used, whether for loading or unloading, and the type of material being handled by the equipment must be identified. The operator of the vehicle being loaded or unloaded must be informed whether he/she can stay in the cab of his/her vehicle. The regulation for remaining in a vehicle being loaded is 1926.601(b)(6), which states "All haulage vehicles, whose pay load is loaded by means of cranes, power shovels, loaders, or similar equipment, shall have a cab shield and/or canopy adequate to protect the operator from shifting or falling materials." When haulage vehicles without a cab shield and/or canopy are being loaded, the operator shall not be permitted to remain in the vehicle while it is being loaded.

(f) Warning System for Mobile Equipment

When mobile equipment is operated adjacent to an excavation, or when such equipment is required to approach the edge of an excavation, and the operator does not have a clear and direct view of the edge of the excavation, a warning system shall be utilized such as barricades, hand or mechanical signals, or stop logs. If possible, the grade should be away from the excavation.

A warning system must utilized if the mobile equipment operator does not have a clear and direct view of the excavation. This system should identify what can obstruct the operator's view and the controls put in place to ensure that the warning system used will keep the operator away from the excavation.

(g) Hazardous Atmospheres

OSHA provided the following when asked if their Confined Spaces Advisor on the Internet applies to construction, and if it does not apply, what section or sections must be followed.

OSHA's Confined Spaces Advisor does not apply to confined spaces in construction. It provides guidance regarding OSHA's permit-required confined space (PRCS) rules for general industry employers under 29 CFR 1910.146. Under §1910.146(a), employers of workers engaged exclusively in construction are not covered by this standard and the OSHA construction industry standards do not require permits for confined space entry.

In construction, employers must comply with §1926.21(b)(6)(i), which states: "All employees required to enter a confined or enclosed space shall be instructed as to the nature of the hazards involved, the necessary precautions to be taken, and in the use of protective and emergency equipment required. The employer shall comply with any specific regulations that apply to the work in dangerous or potentially dangerous areas."

Other regulations that may apply include:

§1926.353(b)(1) (requirement for exhaust ventilation when welding, cutting, or heating is performed in confined spaces);

§1926.353(b)(2) (requirements for air line respirators and standby person whenever the means of access is blocked by ventilation equipment);

§1926.800 Subpart S - Underground Construction (requirements applicable to caissons, etc.);

§1926.956 underground lines (electrical); and

§1926.651(g) (hazardous atmospheres in excavations).

(g)(1)

Testing and controls. In addition to the requirements set forth in subparts D and E of this part (29 CFR 1926.50 - 1926.107) to prevent exposure to harmful levels of atmospheric contaminants and to assure acceptable atmospheric conditions, the following requirements shall apply:

(g)(1)(i)

Where oxygen deficiency (atmospheres containing less than 19.5 percent oxygen) or a hazardous atmosphere exists or could reasonably be expected to exist, such as in excavations in landfill areas or excavations in areas where hazardous substances are stored nearby, the atmospheres in the excavation shall be tested before employees enter excavations greater than 4 feet (1.22 m).

Air sampling required by this section does not have to be performed in all trenches over 4 feet in depth. This paragraph addresses excavations over 4 feet deep where oxygen deficiency or other "hazardous atmosphere exists or could reasonably be expected to exist." These conditions may exist where there is operating equipment in or around the excavation, or when excavations are in landfill areas or in areas where hazardous substances are stored nearby. It is the responsibility of a designated competent person to determine if air monitoring needs to be performed before employees enter each excavation. Atmospheric testing should be conducted as often as necessary to ensure that the atmosphere remains safe to prevent employee exposure to hazardous atmospheric conditions. Only trained personnel should use air monitoring equipment.

When air monitoring is performed, there is no requirement to document the results but it a best practice to do so. The Air Monitoring Data Sheet is an example of ways to document the results of any air monitoring. By using this data sheet, specific operation/conditions (hazards and controls, test results, excavation depth, etc.) can be documented

(g)(1)(ii)

Adequate precautions shall be taken to prevent employee exposure to atmospheres containing less than 19.5 percent oxygen and other hazardous atmospheres. These precautions include providing proper respiratory protection or ventilation in accordance with subparts D and E of this part respectively.

Sewers and excavations containing sewers are normally considered as locations that may contain hazardous atmospheres. Both the Permit-Required

Confined Spaces standard (29 CFR 1910.146) and Subpart P—Excavations in the Construction standard recognize sewers and excavations containing sewers as areas where adequate precautions are to be taken to prevent employee exposure to flammable gases. The general requirements for excavation require that adequate precautions are taken to prevent employee exposure to an atmosphere containing a concentration of a flammable gas in excess of 20 percent (29 CFR 1926.651(g)(1)(iii)). The Permit-Required Confined Spaces standard includes "flammable gas, vapor, or mist in excess of 10 percent of its lower flammable limit" and "airborne combustible dust at a concentration that meets or exceeds its LFL" in the definition of a hazardous atmosphere.

The failure to use equipment and associated wiring approved for the hazardous location could increase the likelihood of an accident. Intrinsically safe equipment and associated wiring approved as intrinsically safe shall be permitted in any hazardous (classified) location for which it is approved (29 CFR 1910.307).

(g)(1)(iii)

Adequate precaution shall be taken such as providing ventilation, to prevent employee exposure to an atmosphere containing a concentration of a flammable gas in excess of 20 percent of the lower flammable limit of the gas.

Describe any hazards observed, including the type and percent of gas identified over the lower limit and the controls implemented.

(g)(1)(iv)

When controls are used that are intended to reduce the level of atmospheric contaminants to acceptable levels, testing shall be conducted as often as necessary to ensure that the atmosphere remains safe.

When testing is required and conducted, readings (results) should be recorded prior to the first excavation entry and when re-entered after breaks, lunch, etc.

(g)(2)

Emergency rescue equipment.

(g)(2)(i)

Emergency rescue equipment, such as breathing apparatus, a safety harness and line, or a basket stretcher, shall be readily available where hazardous atmospheric conditions exist or may reasonably be expected to develop during work in an excavation. This equipment shall be attended when in use.

When a competent person determines, based on the conditions at each jobsite, that hazardous atmospheric conditions exist or may reasonably be expected to develop during work in an excavation, emergency rescue equipment must be readily available and attended when in use. Simply having the emergency rescue equipment on site does not satisfy the requirement of this section; it must be attended if it is on site because of hazardous conditions. As to whether a contractor can rely on a local rescue squad instead of providing the rescue equipment, many emergency situations associated with the risks involved with hazardous atmospheres in trenches would normally require an immediate response within a few minutes or even seconds. A rescue squad would be unable to provide the necessary response and therefore could not be used to comply with this section.

(g)(2)(ii)

Employees entering bell-bottom pier holes, or other similar deep and confined footing excavations, shall wear a harness with a lifeline securely attached to it. The lifeline shall be separate from any line used to handle materials, and shall be individually attended at all times while the employee wearing the lifeline is in the excavation.

In response to a memorandum requesting an interpretation of the requirements addressing the use of cranes to lower employees into drilled pier foundations, OSHA's position was stated as follows: "A boatswain chair suspended by reliable hoisting equipment utilizing an A-frame can be used by employees entering drilled pier foundations provided the applicable standards are met including the procedures outlined in 29 CFR 1926.451, 1926.452(o) and 1926.651(g)(2)(ii). However, if jobsite conditions prevent the use of A-frame equipment or the use of such equipment would be more hazardous than the use of a crane, a mobile crane can be used for this purpose provided the employees are lowered in a personnel platform in accordance with the requirements of 29 CFR 1926.550(g)."

OSHA is aware that a standard personnel platform is often too poorly designed to function well in a drilled pier application. However, they believe a platform could readily be designed that would comply with the requirements of 29 CFR 1926.550(g)(4) and that would both provide protection for the employee entering the drilled pier and allow the necessary work to be performed.

(h) Protection from hazards associated with water accumulation

(h)(1)

Employees shall not work in excavations in which there is accumulated water, or in excavations in which water is accumulating, unless adequate precautions have been taken to protect employees against the hazards posed by water accumulation. The precautions necessary to protect employees adequately vary with each situation, but could include special support or shield systems to protect from cave-ins, water removal to control the level of accumulating water, or use of a safety harness and lifeline.

Workers can enter a trench with water accumulation if the workers are protected from cave-in by shoring, shields, or sloping and the water level is controlled. This section (.651(h)) allows workers to work in a trench with water accumulation, provided adequate precautions have been taken to protect employees against the hazards posed by water accumulation. The precautions necessary to comply with the standard vary with each situation, and precautions such as additional shoring and control of the water level may not, in all cases, provide the required employee protection.

(h)(2)

If water is controlled or prevented from accumulating by the use of water removal equipment, the water removal equipment and operations shall be monitored by a competent person to ensure proper operation.

In order to be a "competent person" for the purpose of water removal equipment, one must have specific training in and be knowledgeable about the water removal equipment used and requirements of the excavation standard.

(h)(3)

If excavation work interrupts the natural drainage of surface water (such as streams), diversion ditches, dikes, or other suitable means shall be used to prevent surface water from entering the excavation and to provide adequate drainage of the area adjacent to the excavation. Excavations subject to runoff from heavy rains will require an inspection by a competent person and compliance with paragraphs (h)(1) and (h)(2) of this section.

An Excavation Condition Log (Appendix A) provides a way to document the date and time of the last inspection by a competent person and, if any, the last heavy rain.

(i) Stability of Adjacent Structures

(i)(1)

Where the stability of adjoining buildings, walls, or other structures is endangered by excavation operations, support systems such as shoring, bracing, or underpinning shall be provided to ensure the stability of such structures for the protection of employees.

A competent/qualified person is required to establish a support system to ensure the protection of employees by stabilizing adjoining structures endangered by any excavation activity.

(i)(2)

Excavation below the level of the base or footing of any foundation or retaining wall that could be reasonably expected to pose a hazard to employees shall not be permitted except when:

(i)(2)(i)

A support system, such as underpinning, is provided to ensure the safety of employees and the stability of the structure; or

(i)(2)(ii)

The excavation is in stable rock; or

(i)(2)(iii)

A registered professional engineer has approved the determination that the structure is sufficiently removed from the excavation so as to be unaffected by the excavation activity; or

(i)(2)(iv)

A registered professional engineer has approved the determination that such excavation work will not pose a hazard to employees.

Any excavation below the level of the base or footing of any foundation or retaining wall that could be reasonably expected to pose a hazard to employees can proceed only after a support system has been designed and approved by a competent person or a registered professional engineer has given signed approval to proceed.

In situations where it is reasonably expected to pose a hazard, registered professional engineer approval is not required when a support system, such as underpinning, is provided to ensure the safety of employees and the stability of the structure, or when the excavation is in stable rock.

(i)(3)

Sidewalks, pavements and appurtenant structure shall not be undermined unless a support system or another method of protection is provided to protect employees from the possible collapse of such structures.

Anytime a structure is undermined and needs a support system or other method of protection to protect employees, describe the structure needing support and the support system provided.

(j) Protection of Employees From Loose Rock or Soil.

(j)(1)

Adequate protection shall be provided to protect employees from loose rock or soil that could pose a hazard by falling or rolling from an excavation face. Such protection shall consist of scaling to remove loose material; installation of protective barricades at intervals as necessary on the face to stop and contain falling material; or other means that provide equivalent protection.

When the possibility exists for loose rock or soil to fall or roll into excavations, employees must be protected as necessary. Documented evaluations by a competent person should be made before employees are permitted to enter the excavation for the first time and at other times as deemed necessary.

(j)(2)

Employees shall be protected from excavated or other materials or equipment that could pose a hazard by falling or rolling into excavations. Protection shall be provided by placing and keeping such materials or equipment at least 2 feet (.61 m) from the edge of excavations, or by the use of retaining devices that are sufficient to prevent materials or equipment from falling or rolling into excavations, or by a combination of both if necessary.

(k) Inspections

(k)(1)

Daily inspections of excavations, the adjacent areas, and protective systems shall be made by a competent person for evidence of a situation that could result in

Figure 6.2. Protection from Loose Rock or Soil.

possible cave-ins, indications of failure of protective systems, hazardous atmospheres, or other hazardous conditions. An inspection shall be conducted by the competent person prior to the start of work and as needed throughout the shift. Inspections shall also be made after every rainstorm or other hazard increasing occurrence. These inspections are only required when employee exposure can be reasonably anticipated.

29 CFR 1926.651(k) requires that the competent person shall conduct an inspection prior to the start of work and as needed throughout the shift. A competent person shall make daily inspections of the excavation, adjacent areas, and protective systems. In situations where no competent person is determined to be available to conduct the daily inspections, OSHA will cite 29 CFR 1926.651(k). Failure to conduct daily inspections of the trench by a competent person and failure to provide an adequate protective system for employees working in a trench is a willful violation.

During initial opening of an excavation and where the job moves through changing types of soil, the competent person must determine the respective types of soil as the job progresses. In order to correctly ascertain the soil types, the competent person must identify the locations and the limits of each type of soil, and must conduct visual and all appropriate manual tests to classify the initial (opening) soil types observed.

In the event that the employer permitted employees to enter an excavation that was alleged to comply with CFR 1926.652(b) or (c), a violation of the standard exists if no competent person is determined to be available. In such an instance, both 29 CFR 1926.651(k) and .652(b) or (c) would be cited and grouped as appropriate.

As to whether the competent person mentioned in 29 CFR 1926.651(k)(1) must be on site at all times, there is no blanket requirement that a competent person be present at a construction jobsite at all times. The competent person can leave the site periodically, consistent with the goals of this section. It is the responsibility of the competent person to make those inspections necessary to identify situations that could result in hazardous conditions (e.g., possible cave-ins, indications of failure of protective systems, hazardous atmospheres, or other hazardous conditions), and then to ensure that corrective measures are taken. It is, therefore, subject to the conditions present at each individual worksite whether or not a competent person is required to be present at the jobsite at all times.

(k)(2)

Where the competent person finds evidence of a situation that could result in a possible cave-in, indications of failure of protective systems, hazardous atmospheres, or other hazardous conditions, exposed employees shall be removed from the hazardous area until the necessary precautions have been taken to ensure their safety.

(l) Fall Protection

(l)(1)

Walkways shall be provided where employees or equipment are required or permitted to cross over excavations. Guardrails which comply with 1926.502(b) shall be provided where walkways are 6 feet (1.8 m) or more above lower levels.

Before employees are permitted to cross over excavations, a competent person is required to provide overall measurements (including the potential fall distance) and make a determination if guardrails are required. When guardrails are provided, the design must comply with the requirements found in 1926.502 (b).

1926.502(b) contains the following requirements:

"Guardrail systems." Guardrail systems and their use shall comply with the following provisions:

(b)(1)

Top edge height of top rails, or equivalent guardrail system members, shall be 42 inches (1.1 m) plus or minus 3 inches (8 cm) above the walking/working level. When conditions warrant, the height of the top edge may exceed the 45-inch height, provided the guardrail system meets all other criteria of this paragraph.

Note: When employees are using stilts, the top edge height of the top rail, or equivalent member, shall be increased an amount equal to the height of the stilts.

(b)(2)

Midrails, screens, mesh, intermediate vertical members, or equivalent intermediate structural members shall be installed between the top edge of the guardrail system and the walking/working surface when there is no wall or parapet wall at least 21 inches (53 cm) high.

(b)(2)(i)

Midrails, when used, shall be installed at a height midway between the top edge of the guardrail system and the walking/working level.

(b)(2)(ii)

Screens and mesh, when used, shall extend from the top rail to the walking/working level and along the entire opening between top rail supports.

(b)(2)(iii)

Intermediate members (such as balusters), when used between posts, shall be not more than 19 inches (48 cm) apart.

(b)(2)(iv)

Other structural members (such as additional midrails and architectural panels) shall be installed such that there are no openings in the guardrail system that are more than 19 inches (.5 m) wide.

(b)(3)

Guardrail systems shall be capable of withstanding, without failure, a force of at least 200 pounds (890 N) applied within 2 inches (5.1 cm) of the top edge, in any outward or downward direction, at any point along the top edge.

(b)(4)

When the 200 pound (890 N) test load specified in paragraph (b)(3) of this section is applied in a downward direction, the top edge of the guardrail shall not deflect to a height less than 39 inches (1.0 m) above the walking/working level. Guardrail system components selected and constructed in accordance with the Appendix B to subpart M of this part will be deemed to meet this requirement.

(b)(5)

Midrails, screens, mesh, intermediate vertical members, solid panels, and equivalent structural members shall be capable of withstanding, without failure, a force of at least 150 pounds (666 N) applied in any downward or outward direction at any point along the midrail or other member.

(b)(6)

Guardrail systems shall be so surfaced as to prevent injury to an employee from punctures or lacerations, and to prevent snagging of clothing.

(b)(7)

The ends of all top rails and midrails shall not overhang the terminal posts, except where such overhang does not constitute a projection hazard.

(b)(8)

Steel banding and plastic banding shall not be used as top rails or midrails.

(b)(9)

Top rails and midrails shall be at least one-quarter inch (0.6 cm) nominal diameter or thickness to prevent cuts and lacerations. If wire rope is used for top rails, it shall be flagged at not more than 6-foot intervals with high-visibility material.

(b)(10)

When guardrail systems are used at hoisting areas, a chain, gate or removable guardrail section shall be placed across the access opening between guardrail sections when hoisting operations are not taking place.

(b)(11)

When guardrail systems are used at holes, they shall be erected on all unprotected sides or edges of the hole.

(b)(12)

When guardrail systems are used around holes used for the passage of materials, the hole shall have not more than two sides provided with removable guardrail sections to allow the passage of materials. When the hole is not in use, it shall be closed over with a cover, or a guardrail system shall be provided along all unprotected sides or edges.

(b)(13)

When guardrail systems are used around holes which are used as points of access (such as ladderways), they shall be provided with a gate, or be so offset that a person cannot walk directly into the hole.

(b)(14)

Guardrail systems used on ramps and runways shall be erected along each unprotected side or edge.

(b)(15)

Manila, plastic or synthetic rope being used for top rails or midrails shall be inspected as frequently as necessary to ensure that it continues to meet the strength requirements of paragraph (b)(3) of this section.

4.C. – 29 CFR 1926.652 – Requirements for Protective Systems.

- Part Number: 1926
- Part Title: Safety and Health Regulations for Construction
- Subpart : P
- Subpart Title: Excavations
- Standard Number: 1926.652
- Title: Requirements for Protective Systems.

(a) Protection of Employees in Excavations.

On June 30, 1995, OSHA issued a memorandum that in part stated "Effective immediately and until further notice, 29 CFR 1926.652 shall not be applied to house foundation/basement excavations when all the following conditions are present."

However, where the excavation walls are used as the form for the gunite or shotcrete pool walls to swimming pool excavations, compliance with 1926.652(a) has been established as being infeasible for this type of pool construction, but all remaining requirements of Part 1926, Subpart P still apply.

This was established after members of the Office of Construction and Maritime Compliance Assistance, Office of Construction and Engineering, and the Office of Construction and Civil Engineering Safety standards researched this issue. These offices determined that compliance with the provisions of Subpart P during the construction of gunite or shotcrete pool walls is not feasible. They concluded that application of 1926.652(a) requiring a protective system to be provided in swimming pool excavations would prevent the performance of the required work.

In instances where it has been established that compliance with a standard is infeasible, it is OSHA's policy to require the employer to use alternative means or methods of avoiding or reducing the hazard. (OSHA Instruction CPL 2.45B CH-3, Chapter V section E.3.b.) Such alternative means and methods for swimming pool excavations include at least the following:

1) An effective thorough safety program is in place.

2) Employees are thoroughly trained in the provisions of the safety plan.

3) All spoil piles are maintained a safe distance away from excavation (equal to or greater distance than depth).

4) Vehicles are kept a safe distance away from excavation (equal to or greater distance than depth).

5) All portions of the excavation that appear to the competent person to pose a hazard of sloughing or cave-in shall be removed or stabilized.

6) A competent person is present at all times while employees are exposed to cave-in hazards.

7) A protective system must be provided on all non-form excavation walls over 5 feet high.

(a)(1)

Each employee in an excavation shall be protected from cave-ins by an adequate protective system designed in accordance with paragraph (b) or (c) of this section except when:

(a)(1)(i)

Excavations are made entirely in stable rock; or

(a)(1)(ii)

Excavations are less than 5 feet (1.52 m) in depth and examination of the ground by a competent person provides no indication of a potential cave-in.

Each employee in an excavation must be protected from cave-ins by an adequate protective system designed in accordance with 29 CFR 1926.652 (b) or (c). For sloping to be adequate, the employer must comply with the provisions of 29 CFR 1926.652(b)(1)(i) in that the excavation be sloped at an angle steeper than one and one-half horizontal to one vertical (34 degrees measured from the horizontal). Configurations for benching systems shall be determined in accordance with the conditions and requirements set forth in appendices A and B to this subpart.

(a)(2)

Protective systems shall have the capacity to resist without failure all loads that are intended or could reasonably be expected to be applied or transmitted to the system.

(b) Design of Sloping and Benching Systems

The slopes and configurations of sloping and benching systems shall be selected and constructed by the employer or his designee and shall be in accordance with the requirements of paragraph (b)(1); or, in the alternative, paragraph (b)(2); or, in the alternative, paragraph (b)(3); or, in the alternative, paragraph (b)(4), as follows:

(b)(1)

Option (1) - Allowable configurations and slopes.

(b)(1)(i)

Excavations shall be sloped at an angle not steeper than one and one-half horizontal to one vertical (34 degrees measured from the horizontal), unless the employer uses one of the other options listed below.

(b)(1)(ii)

Slopes specified in paragraph (b)(1)(i) of this section, shall be excavated to form configurations that are in accordance with the slopes shown for Type C soil in Appendix B to this subpart.

Protective systems constructed to the requirements of 29 CFR 1926.652(c), Option (1), must be inspected against the requirements of appendices A, C, or D. The competent person must be thoroughly knowledgeable of those requirements and must inspect daily to assure the continuing integrity of the system.

Employers who elect to comply with the specifications of Option (1) at 29 CFR 1926.652(b)(1) must meet all the necessary requirements of the standard, except that a competent person must perform daily inspections as required at .651(k). Such inspections can be anticipated to primarily address hazardous atmospheres or localized problems related to rainstorms and/or changing weather conditions.

(b)(2)

Option (2) - Determination of slopes and configurations using Appendices A and B. Maximum allowable slopes, and allowable configurations for sloping and benching systems, shall be determined in accordance with the conditions and requirements set forth in appendices A and B to this subpart.

Protective systems constructed to Option (2) of 29 CFR 1926.652(c) must be inspected against the manufacturer's specifications, recommendations, and limitations for the system and it must be verified that any deviation, if issued, continues to meet the manufacturer's specific written approval. The competent person must be thoroughly knowledgeable of the limitations of the manufactured system.

Employers who elect to comply with the requirements of Options (2) through (4) at 29 CFR 1926.652(b)(2), (3), and (4), must conduct, with the services of a competent person, the initial and daily visual inspections together with appropriate manual tests of the soil along the excavation. The slope, shoring, or shield system used must reflect the results of such inspections and tests if any of the soil is determined to be Type A, B, or equivalent as designated by the registered professional engineer. Visual observation and at least random manual testing is required if no visual changes are noted. Type C soil need not be manually tested once it is visually determined to be Type C and an appropriate protective system is in place. On the other hand, Type A or B soil must be manually tested to verify its type. During the daily or other inspection, required at .651(k), the competent person must visually ascertain if soil changes are occurring that could lower the classification of the previously categorized soil type. If, in the competent person's opinion (which is based upon the visual criteria of the standard), deteriorating soil changes are occurring or have occurred, manual testing is mandatory and the soil classification must be ascertained. For instance, where Type A soil has deteriorated to Type B, based upon visual observation, manual testing must be conducted to confirm the condition of the soil. When the soil is downgraded to Type C and a corrective protective system is put in place, no manual testing is required or necessary. It must be realized that once an excavation is opened, the soil type observed to exist at the time of opening will not improve with time; it can only deteriorate.

Designs Using Manufacturer's Tabulated Data, set forth in 1926.652(c)(2), allows the use of tables other than those given in the OSHA standards. However, by definition, "tabulated data" means "tables and charts ***approved by a registered professional engineer*** *and used to design and construct a protective system" (emphasis added). Therefore, the C-60 soil classification may be used in the design of protective systems if the classification and subsequent designs have been approved by a registered professional* *engineer. Manufacturer's tabulated data can be used to design protective systems for trenches more than 20 feet in depth provided the use is within the limits of the data, including depth limitations and soil type.*

As to whether a contractor must have separate manufacturer's tabulated data on hand for each specific shield, only one set of tabulated data is required for each different shield design. If a contractor uses several shields of identical make and model, only one set of tabulated data would be required for them.

(b)(3)

Option (3) - Designs using other tabulated data.

(b)(3)(i)

Designs of sloping or benching systems shall be selected from and in accordance with tabulated data, such as tables and charts.

(b)(3)(ii)

The tabulated data shall be in written form and shall include all of the following:

(b)(3)(ii)(A)

Identification of the parameters that affect the selection of a sloping or benching system drawn from such data;

(b)(3)(ii)(B)

Identification of the limits of use of the data, to include the magnitude and configuration of slopes determined to be safe;

(b)(3)(ii)(C)

Explanatory information as may be necessary to aid the user in making a correct selection of a protective system from the data.

(b)(3)(iii)

At least one copy of the tabulated data which identifies the registered professional engineer who approved the data, shall be maintained at the jobsite during construction of the protective system. After that time the data may be stored off the jobsite, but a copy of the data shall be made available to the Secretary upon request.

Protective systems constructed to the Option (3) requirement of 29 CFR 1926.652(c) must meet the stipulations of the tabulated data, which must be approved by a registered professional engineer. The competent person who inspects the system on a daily basis must be able to verify that the system continues to maintain the integrity required by the tabulated data. In the event that local deviations are noted, the competent person should consult with a registered professional engineer who can recommend proper corrective measures.

(b)(4)

Option (4) - Design by a registered professional engineer.

(b)(4)(i)

Sloping and benching systems not utilizing Option (1) or Option (2) or Option (3) under paragraph (b) of this section shall be approved by a registered professional engineer.

(b)(4)(ii)

Designs shall be in written form and shall include at least the following:

(b)(4)(ii)(A)

The magnitude of the slopes that were determined to be safe for the particular project;

(b)(4)(ii)(B)

The configurations that were determined to be safe for the particular project;

(b)(4)(ii)(C)

The identity of the registered professional engineer approving the design.

(b)(4)(iii)

At least one copy of the design shall be maintained at the jobsite while the slope is being constructed. After that time the design need not be at the jobsite, but a copy shall be made available to the Secretary upon request.

Option (4) of 29 CFR 1926.652(c) requires that a registered professional engineer design the protective system. The competent person who inspects the system on a daily basis must be knowledgeable of the system design and its limitations. The competent person should report all deviations noted during inspections to the registered professional engineer so that proper corrective measures can be taken.

Subpart P of 29 CFR 1926 allows an employer four options when selecting or designing a system to protect employees working in excavations. To use Option (1), the soil type (A, B, or C) at each excavation must be determined as outlined in Appendix A of Subpart P. The type A, B, and C soil classification system is applicable only when designing a system under Option (1) (use of the tables contained in appendices C and D.) An employer is not bound by the A, B, and C soil classification if the employer elects to provide a protective system under Options (2), (3), or (4).

The soil at each excavation must be tested for both type and weight. If the employer elects to use Option (1), the soil must be assumed to be type C or must be tested at every excavation in accordance with Appendix A. If a protective system is designed under Options (2), (3), or (4), the employer has the responsibility to make whatever tests are necessary to ascertain all criteria and limitations that a manufacturer or designer (as approved by a registered professional engineer) specifies or recommends regarding the use of the protective system, and to assure that the system is used accordingly. This would include a determination that the limitations of the system are not exceeded by the soil at each excavation.

(c) Design of Support Systems, Shield Systems, and Other Protective Systems.

Designs of support systems, shield systems, and other protective systems shall be selected and constructed by the employer or his designee and shall be in accordance with the requirements of paragraph (c)(1); or, in the alternative, paragraph (c)(2); or, in the alternative, paragraph (c)(3); or, i the alternative, paragraph (c)(4) as follows:

(c)(1)

Option (1) - Designs using appendices A, C and D. Designs for timber shoring in trenches shall be determined in accordance with the conditions and requirements set forth in appendices A and C to this subpart. Designs for aluminum hydraulic shoring shall be in accordance with paragraph (c)(2) of this section, but if manufacturer's tabulated data cannot be utilized, designs shall be in accordance with appendix D.

There are instances where two layers of 3/4-inch thick plywood can be used to hold back raveling when used with an aluminum hydraulic waler system. Where an employer chooses to design a protective system in accordance with 29 CFR 1926.652(c)(1), Option (1), plywood cannot be used as a substitute for any primary load carrying structural members such as the timber uprights or aluminum rails as called for in Tables D-1.1 through D-1.4. However, plywood can be used to contain raveled soil and to transfer the loads imposed by this soil to the load carrying members.

(c)(2)

Option (2) - Designs Using Manufacturer's Tabulated Data.

(c)(2)(i)

Design of support systems, shield systems, or other protective systems that are drawn from manufacturer's tabulated data shall be in accordance with all specifications, recommendations, and limitations issued or made by the manufacturer.

(c)(2)(ii)

Deviation from the specifications, recommendations, and limitations issued or made by the manufacturer shall only be allowed after the manufacturer issues specific written approval.

(c)(2)(iii)

Manufacturer's specifications, recommendations, and limitations, and manufacturer's approval to deviate from the specifications, recommendations, and limitations shall be in written form at the jobsite during construction of the protective system. After that time this data may be stored off the jobsite, but a copy shall be made available to the Secretary upon request.

Option (2) allows support systems, shield systems, and other protective systems to be designed using manufacturer's tabulated data without additional certification from a registered professional engineer. Tabulated data by definition (1926.650(b)) is required to be certified by a registered professional engineer and additional certification is not necessary.

(c)(3) Option (3) - Designs using other tabulated data.

An employer who elects to use Option (3) must meet only the requirements of this subsection and is not bound by the requirements in the appendices of 1926.652. Registered professional engineers who develop tabulated data for use by employers under Option (3) often will include plywood sheeting as a component of support systems. To aid such engineers, it may be worthwhile to provide data on the equivalency of various combinations of commonly available plywood types and sizes.

(c)(3)(i)

Designs of support systems, shield systems, or other protective systems shall be selected from and be in accordance with tabulated data, such as tables and charts.

(c)(3)(ii)

The tabulated data shall be in written form and include all of the following:

(c)(3)(ii)(A)

Identification of the parameters that affect the selection of a protective system drawn from such data;

(c)(3)(ii)(B)

Identification of the limits of use of the data;

(c)(3)(ii)(C)

Explanatory information as may be necessary to aid the user in making a correct selection of a protective system from the data.

(c)(3)(iii)

At least one copy of the tabulated data, which identifies the registered professional engineer who approved the data, shall be maintained at the jobsite during construction of the protective system. After that time the data may be stored off the jobsite, but a copy of the data shall be made available to the Secretary upon request.

Protective systems for excavations that are in compliance with the requirements of 29 CFR 1926.652(c) must be inspected daily to verify that the support systems installed remain effective. The knowledge and experience of the competent person needed under this requirement are therefore considerably greater than that required of a competent person under the requirements of 29 CFR 1926.652(b).

The question often arises whether OSHA requires manufacturer's tabulated data (MTD) to identify a registered professional engineer. The OSHA regulations at 1926.652(c) Design of Support Systems, Shield Systems, and Other Protective Systems, set forth four options for the design of protective trench systems. Option (3) - Designs using other tabulated data, and Option (4) - Design by a registered professional engineer, respectively require the registered professional engineer who develops the tabulated data or approves the design of the protective system to be identified in writing and such information must be maintained at the jobsite. However, the other options, including Option (2) - Designs Using Manufacturer's Tabulated Data, do not specifically require the registered engineer to be identified in writing.

Although Option (2) does not specifically require the engineer who prepared the MTD to be identified, MTD by definition means that tables and charts are approved by a registered professional engineer. Without proof of this approval, such as an engineer's stamp, it will be difficult for a contractor using an MTD-based system to establish compliance with the requirements of Subpart P. Therefore, it may be of benefit to provide this information although it is not required.

(c)(4)

Option (4) - Design by a registered professional engineer.

(c)(4)(i)

Support systems, shield systems, and other protective systems not utilizing Option 1, Option 2 or Option 3, above, shall be approved by a registered professional engineer.

(c)(4)(ii)

Designs shall be in written form and shall include the following:

(c)(4)(ii)(A)

A plan indicating the sizes, types, and configurations of the materials to be used in the protective system; and

(c)(4)(ii)(B)

The identify of the registered professional engineer approving the design.

(c)(4)(iii)

At least one copy of the design shall be maintained at the jobsite during construction of the protective system. After that time, the design may be stored off the jobsite, but a copy of the design shall be made available to the Secretary upon request.

(d) Materials and Equipment.

(d)(1)

Materials and equipment used for protective systems shall be free from damage or defects that might impair their proper function.

(d)(2)

Manufactured materials and equipment used for protective systems shall be used and maintained in a manner that is consistent with the recommendations of the manufacturer, and in a manner that will prevent employee exposure to hazards.

(d)(3)

When material or equipment that is used for protective systems is damaged, a competent person shall examine the material or equipment and evaluate its suitability for continued use. If the competent person cannot assure the material or equipment is able to support the intended loads or is otherwise suitable for safe use, then such material or equipment shall be removed from service, and shall be evaluated and approved by a registered professional engineer before being returned to service.

(e) Installation and Removal of Support

(e)(1)

General.

(c)(1)(i)

Members of support systems shall be securely connected together to prevent sliding, falling, kickouts, or other predictable failure.

(e)(1)(ii)

Support systems shall be installed and removed in a manner that protects employees from cave-ins, structural collapses, or from being struck by members of the support system.

(e)(1)(iii)

Individual members of support systems shall not be subjected to loads exceeding those which those members were designed to withstand.

(e)(1)(iv)

Before temporary removal of individual members begins, additional precautions shall be taken to ensure the safety of employees, such as installing other structural members to carry the loads imposed on the support system.

(e)(1)(v)

Removal shall begin at, and progress from, the bottom of the excavation. Members shall be released slowly so as to note any indication of possible failure of the remaining members of the structure or possible cave-in of the sides of the excavation.

(e)(1)(vi)

Backfilling shall progress together with the removal of support systems from excavations.

(e)(2)

Additional requirements for support systems for trench excavations.

(e)(2)(i)

Excavation of material to a level no greater than 2 feet (.61 m) below the bottom of the members of a support system shall be permitted, but only if the system is designed to resist the forces calculated for the full depth of the trench, and there are no indications while the trench is open of a possible loss of soil from behind or below the bottom of the support system.

When trench shields are stacked in a deep trench, they do not all have to be designed to resist the forces calculated for the deepest part of the trench. The standard only requires that the shields be used at a depth not greater than that for which they are designed. To facilitate proper stacking in deep excavations, a design or system should be used to prevent the shields from being misplaced in the trench when shields of different ratings are used. Designing the shields so they will fit together only one correct way, or using a marking system that readily identifies the maximum depth each individual shield can be placed are examples of systems that could be used to prevent such misplacement.

(e)(2)(ii)

Installation of a support system shall be closely coordinated with the excavation of trenches.

(f) Sloping and benching systems

Employees shall not be permitted to work on the faces of sloped or benched excavations at levels above other employees except when employees at the lower levels are adequately protected from the hazard of falling, rolling, or sliding material or equipment.

(g) Shield systems

(g)(1) General.

(g)(1)(i)

Shield systems shall not be subjected to loads exceeding those which the system was designed to withstand.

(g)(1)(ii)

Shields shall be installed in a manner to restrict lateral or other hazardous movement of the shield in the event of the application of sudden lateral loads.

(g)(1)(iii)

Employees shall be protected from the hazard of cave-ins when entering or exiting the areas protected by shields.

(g)(1)(iv)

Employees shall not be allowed in shields when shields are being installed, removed, or moved vertically.

(g)(2)

Additional requirement for shield systems used in trench excavations. Excavations of earth material to a level not greater than 2 feet (.61 m) below the bottom of a shield shall be permitted, but only if the shield is designed to resist the forces calculated for the full depth of the trench, and there are no indications while the trench is open of a possible loss of soil from behind or below the bottom of the shield.

4.D. – 29 CFR 1926 Subpart P Appendix A - Soil Classification

- Part Number: 1926
- Part Title: Safety and Health Regulations for Construction
- Subpart: P
- Subpart Title: Excavations
- Standard Number: 1926 Subpart P App A
- Title: Soil Classification

(a) Scope and application

(1) Scope. This appendix describes a method of classifying soil and rock deposits based on site and environmental conditions, and on the structure and composition of the earth deposits. The appendix contains definitions, sets forth requirements, and describes acceptable visual and manual tests for use in classifying soils.

(2) Application. This appendix applies when a sloping or benching system is designed in accordance with the requirements set forth in 1926.652(b)(2) as a method of protection for employees from cave-ins. This appendix also applies when timber shoring for excavations is designed as a method of protection from cave-ins in accordance with appendix C to subpart P of part 1926, and when aluminum hydraulic shoring is designed in accordance with appendix D. This Appendix also applies if other protective systems are designed and selected for use from data prepared in accordance with the requirements set forth in 1926.652(c), and the use of the data is predicated on the use of the soil classification system set forth in this appendix.

The question is often asked whether distance alone is adequate to protect employees from the hazards of cave-ins. When employees maintain a distance of at least two times the height of a vertical sidewall from the toe of the sidewall, the employees are not exposed to the hazards of a cave-in. Therefore, no additional protective system is necessary. However, when distance alone is relied upon to protect employees from the hazards of cave-in, all employees entering the excavation must be instructed not to enter the danger zone created by the vertical face. A warning system must be provided to prevent workers from entering the danger zone inadvertently. Roping off the area or adequately marking the area with cones, flags or other highly visible means are examples of acceptable warning systems. It is the responsibility of a competent person to determine and maintain, based on the conditions existing at each site, the distance away from the highwall necessary to protect the workers and the warning system necessary to adequately warn workers of the hazardous area.

(b) Definitions

The definitions and examples given below are based on, in whole or in part, the following; American Society for Testing Materials (ASTM) Standards D653-85 and D2488; The Unified Soils Classification System; The U.S. Department of Agriculture (USDA) Textural Classification Scheme; and The National Bureau of Standards Report BSS-121.

"Cemented soil" means a soil in which the particles are held together by a chemical agent, such as calcium carbonate, such that a hand-size sample cannot be crushed into powder or individual soil particles by finger pressure.

"Cohesive soil" means clay (fine grained soil), or soil with a high clay content, which has cohesive strength. Cohesive soil does not crumble, can be excavated with vertical sideslopes, and is plastic when moist.

Cohesive soil is hard to break up when dry, and exhibits significant cohesion when submerged. Cohesive soils include clayey silt, sandy clay, silty clay, clay and organic clay.

"Dry soil" means soil that does not exhibit visible signs of moisture content.

"Fissured" means a soil material that has a tendency to break along definite planes of fracture with little resistance, or a material that exhibits open cracks, such as tension cracks, in an exposed surface.

"Granular soil" means gravel, sand, or silt (coarse grained soil) with little or no clay content. Granular soil has no cohesive strength. Some moist granular soils exhibit apparent cohesion. Granular soil cannot be molded when moist and crumbles easily when dry.

"Layered system" means two or more distinctly different soil or rock types arranged in layers.

Micaceous seams or weakened planes in rock or shale are considered layered.

"Moist soil" means a condition in which a soil looks and feels damp. Moist cohesive soil can easily be shaped into a ball and rolled into small diameter threads before crumbling. Moist granular soil that contains some cohesive material will exhibit signs of cohesion between particles.

"Plastic" means a property of a soil which allows the soil to be deformed or molded without cracking, or appreciable volume change.

"Saturated soil" means a soil in which the voids are filled with water. Saturation does not require flow. Saturation, or near saturation, is necessary for the proper use of instruments such as a pocket penetrometer or sheer vane.

"Soil classification system" means, for the purpose of this subpart, a method of categorizing soil and rock deposits in a hierarchy of Stable Rock, Type A, Type B, and Type C, in decreasing order of stability. The categories are determined based on an analysis of the properties and performance characteristics of the deposits and the characteristics of the deposits and the environmental conditions of exposure.

"Stable rock" means natural solid mineral matter that can be excavated with vertical sides and remain intact while exposed.

OSHA issued an interpretation in February 1995 concerning whether they should consider organic rock, such as "Miami oolite rock," as "stable rock" for the purposes of OSHA excavation regulations.

The term "stable rock" as defined by OSHA means "natural solid mineral matter that can be excavated with vertical sides and remain intact while exposed." This definition is important because certain precautions that need to be taken to prevent cave-in hazards in other types of material do not need to be taken in stable rock. The concern was that Miami oolite rock is an organic rock and does not meet OSHA's definition and, therefore, does not qualify for the exemption.

After discussing the matter with several soil mechanics experts, OSHA felt it would be inappropriate to change the definition of stable rock to include Miami oolite rock. Although oolite can sometimes exhibit the properties of stable rock, it is often interspersed with sand seams, which are not stable. In addition, oolite rock poses a potential cave-in hazard because of the high angle of fracture that occurs when it is broken, as can happen during excavation operations. This high angle, which is not a characteristic of stable rock, can lead to a cave-in. Because of these characteristics, a blanket exception from the regulations should not be granted for oolite rock conditions. However, OSHA recognizes that protective systems are not always necessary in some soils including oolitic rock, and has provided in the regulations at 29 CFR 1926.652(b)(4) that excavations may proceed in accordance with the recommendations of a registered professional engineer. Therefore, if an engineer evaluates the oolitic rock and determines that protective systems are not necessary, work may proceed accordingly.

"Submerged soil" means soil which is underwater or is free seeping.

"Type A" means:

Cohesive soils with an unconfined, compressive strength of 1.5 ton per square foot (tsf) (144 kPa) or greater. Examples of cohesive soils are: clay, silty clay, sandy clay, clay loam and, in some cases, silty clay loam and sandy clay loam. Cemented soils such as caliche and hardpan are also considered Type A. However, no soil is Type A if:

(i) The soil is fissured; or

(ii) The soil is subject to vibration from heavy traffic, pile driving, or similar effects; or

(iii) The soil has been previously disturbed; or

(iv) The soil is part of a sloped, layered system where the layers dip into the excavation on a slope of four horizontal to one vertical (4H:1V) or greater; or

(v) The material is subject to other factors that would require it to be classified as a less stable material.

"Type B" soil means:

(i) Cohesive soil with an unconfined compressive strength greater than 0.5 tsf (48 kPa) but less than 1.5 tsf (144 kPa); or

(ii) Granular cohesionless soils including: angular gravel (similar to crushed rock), silt, silt loam, sandy loam and, in some cases, silty clay loam and sandy clay loam.

(iii) Previously disturbed soils except those which would otherwise be classed as Type C soil.

(iv) Soil that meets the unconfined compressive strength or cementation requirements for Type A, but is fissured or subject to vibration; or

(v) Dry rock that is not stable; or

(vi) Material that is part of a sloped, layered system where the layers dip into the excavation on a slope less steep than four horizontal to one vertical (4H:1V), but only if the material would otherwise be classified as Type B.

"Type C" soil means:

(i) Cohesive soil with an unconfined compressive strength of 0.5 tsf (48 kPa) or less; or

(ii) Granular soils including gravel, sand, and loamy sand; or

(iii) Submerged soil or soil from which water is freely seeping; or

(iv) Submerged rock that is not stable, or

(v) Material in a sloped, layered system where the layers dip into the excavation or a slope of four horizontal to one vertical (4H:1V) or steeper.

Type C soil is the lowest strength (weakest) soil, and if the employer meets the requirements of 29 C.F.R. 1926.652 (Requirements for Protective Systems) and Appendices A, B, C, and D of Subpart P, as applicable to Type C soil, soil testing would be superfluous and not required by OSHA. However, the employer bears the burden of proof that this is, indeed, the case.

In the event that an employer permitted employees to enter an excavation that was alleged to comply with CFR 1926.652(b) or (c), a violation of the standard exists if no competent person is determined to be available. In such an instance, both 29 CFR 1926.651(k) and .652(b) or (c) would be cited and grouped as appropriate.

"Unconfined compressive strength" means the load per unit area at which a soil will fail in compression. It can be determined by laboratory testing, or estimated in the field using a pocket penetrometer, by thumb penetration tests, and other methods.

"Wet soil" means soil that contains significantly more moisture than moist soil, but in such a range of values that cohesive material will slump or begin to flow when vibrated. Granular material that would exhibit cohesive properties when moist will lose those cohesive properties when wet.

(c) Requirements

(1) Classification of soil and rock deposits. Each soil and rock deposit shall be classified by a competent person as Stable Rock, Type A, Type B, or Type C in accordance with the definitions set forth in paragraph (b) of this appendix.

(2) Basis of classification. The classification of the deposits shall be made based on the results of at least one visual and at least one manual analysis. Such analyses shall be conducted by a competent person using tests described in paragraph (d) below, or in other recognized methods of soil classification and testing such as those adopted by the American Society for Testing Materials, or the U.S. Department of Agriculture textural classification system.

(3) Visual and manual analyses. The visual and manual analyses, such as those noted as being acceptable in paragraph (d) of this appendix, shall be designed and conducted to provide sufficient quantitative and qualitative information as may be necessary to identify properly the properties, factors, and conditions affecting the classification of the deposits.

(4) Layered systems. In a layered system, the system shall be classified in accordance with its weakest layer. However, each layer may be classified individually where a more stable layer lies under a less stable layer.

(5) Reclassification. If, after classifying a deposit, the properties, factors, or conditions affecting its classification change in any way, the changes shall be

evaluated by a competent person. The deposit shall be reclassified as necessary to reflect the changed circumstances.

(d) Acceptable visual and manual tests

(1) Visual tests. Visual analysis is conducted to determine qualitative information regarding the excavation site in general, the soil adjacent to the excavation, the soil forming the sides of the open excavation, and the soil taken as samples from excavated material.

(i) Observe samples of soil that are excavated and soil in the sides of the excavation. Estimate the range of particle sizes and the relative amounts of the particle sizes. Soil that is primarily composed of fine-grained material is cohesive material. Soil composed primarily of coarse-grained sand or gravel is granular material.

(ii) Observe soil as it is excavated. Soil that remains in clumps when excavated is cohesive. Soil that breaks up easily and does not stay in clumps is granular.

(iii) Observe the side of the opened excavation and the surface area adjacent to the excavation. Crack-like openings such as tension cracks could indicate fissured material. If chunks of soil spall off a vertical side, the soil could be fissured. Small spalls are evidence of moving ground and are indications of potentially hazardous situations.

(iv) Observe the area adjacent to the excavation and the excavation itself for evidence of existing utility and other underground structures, and to identify previously disturbed soil.

(v) Observed the opened side of the excavation to identify layered systems. Examine layered systems to identify if the layers slope toward the excavation. Estimate the degree of slope of the layers.

(vi) Observe the area adjacent to the excavation and the sides of the opened excavation for evidence of surface water, water seeping from the sides of the excavation, or the location of the level of the water table.

(vii) Observe the area adjacent to the excavation and the area within the excavation for sources of vibration that may affect the stability of the excavation face.

(2) Manual tests. Manual analysis of soil samples is conducted to determine quantitative as well as qualitative properties of soil and to provide more information in order to classify soil properly.

(i) Plasticity. Mold a moist or wet sample of soil into a ball and attempt to roll it into threads as thin as 1/8-inch in diameter. Cohesive material can be successfully rolled into threads without crumbling. For example, if at least a two inch (50 mm) length of 1/8-inch thread can be held on one end without tearing, the soil is cohesive.

(ii) Dry strength. If the soil is dry and crumbles on its own or with moderate pressure into individual grains or fine powder, it is granular (any combination of gravel, sand, or silt). If the soil is dry and falls into clumps which break up into smaller clumps, but the smaller clumps can only be broken up with difficulty, it may be clay in any combination with gravel, sand or silt. If the dry soil breaks into clumps which do not break up into small clumps and which can only be broken with difficulty, and there is no visual indication the soil is fissured, the soil may be considered unfissured.

(iii) Thumb penetration. The thumb penetration test can be used to estimate the unconfined compressive strength of cohesive soils. This test is based on the thumb penetration test described in American Society for Testing and Materials (ASTM) Standard designation D2488 - "Standard Recommended Practice for Description of Soils (Visual - Manual Procedure).") Type A soils with an unconfined compressive strength of 1.5 tsf can be readily indented by the thumb; however, they can be penetrated by the thumb only with very great effort. Type C soils with an unconfined compressive strength of 0.5 tsf can be easily penetrated several inches by the thumb, and can be molded by light finger pressure. This test should be conducted on an undisturbed soil sample, such as a large clump of spoil, as soon as practicable after excavation to keep to a minimum the effects of exposure to drying influences. If the excavation is later exposed to wetting influences (rain, flooding), the classification of the soil must be changed accordingly.

(iv) Other strength tests. Estimates of unconfined compressive strength of soils can also be obtained by use of a pocket penetrometer or by using a hand-operated shearvane.

(v) Drying test. The basic purpose of the drying test is to differentiate between cohesive material with fissures, unfissured cohesive material, and granular material. The procedure for the drying test involves drying a sample of soil that is approximately one inch thick (2.54 cm) and six inches (15.24 cm) in diameter until it is thoroughly dry:

   (A) If the sample develops cracks as it dries, significant fissures are indicated.

   (B) Samples that dry without cracking are to be broken by hand. If considerable force is necessary to break a sample, the soil has significant cohesive material content. The soil can be classified as an unfissured cohesive material and the unconfined compressive strength should be determined.

(C) If a sample breaks easily by hand, it is either a fissured cohesive material or a granular material. To distinguish between the two, pulverize the dried clumps of the sample by hand or by stepping on them. If the clumps do not pulverize easily, the material is cohesive with fissures. If they pulverize easily into very small fragments, the material is granular.

4.E. – 29 CFR 1926 Subpart P Appendix B – Sloping and Benching

• Part Number:	1926
• Part Title:	Safety and Health Regulations for Construction
• Subpart:	P
• Subpart Title:	Excavations
• Standard Number:	1926 Subpart P App B
• Title:	Sloping and Benching

(a) Scope and Application

This appendix contains specifications for sloping and benching when used as methods of protecting employees working in excavations from cave-ins. The requirements of this appendix apply when the design of sloping and benching protective systems is to be performed in accordance with the requirements set forth in 1926.652(b)(2).

(b) Definitions

"Actual slope" means the slope to which an excavation face is excavated.

"Distress" means that the soil is in a condition where a cave-in is imminent or is likely to occur. Distress is evidenced by such phenomena as the development of fissures in the face of or adjacent to an open excavation; the subsidence of the edge of an excavation; the slumping of material from the face or the bulging or heaving of material from the bottom of an excavation; the spalling of material from the face of an excavation; and ravelling, i.e., small amounts of material such as pebbles or little clumps of material suddenly separating from the face of an excavation and trickling or rolling down into the excavation.

"Maximum allowable slope" means the steepest incline of an excavation face that is acceptable for the most favorable site conditions as protection against cave-ins, and is expressed as the ratio of horizontal distance to vertical rise (H:V).

"Short term exposure" means a period of time less than or equal to 24 hours that an excavation is open.

(c) Requirements

(1) Soil classification. Soil and rock deposits shall be classified in accordance with appendix A to subpart P of part 1926.

(2) Maximum allowable slope. The maximum allowable slope for a soil or rock deposit shall be determined from Table B-1 of this appendix.

(3) Actual slope. (i) The actual slope shall not be steeper than the maximum allowable slope.

(ii) The actual slope shall be less steep than the maximum allowable slope, when there are signs of distress. If that situation occurs, the slope shall be cut back to an actual slope which is at least 1/2 horizontal to one vertical (1/2H:1V) less steep than the maximum allowable slope.

(ii) When surcharge loads from stored material or equipment, operating equipment, or traffic are present, a competent person shall determine the degree to which the actual slope must be reduced below the maximum allowable slope, and shall assure that such reduction is achieved. Surcharge loads from adjacent structures shall be evaluated in accordance with 1926.651(i).

(4) Configurations

Configurations of sloping and benching systems shall be in accordance with Figure B-1.

TABLE B-1
MAXIMUM ALLOWABLE SLOPES

SOIL OR ROCK TYPE	MAXIMUM ALLOWABLE SLOPES (H:V)[1] FOR EXCAVATIONS LESS THAN 20 FEET DEEP[3]
STABLE ROCK	VERTICAL (90 Deg.)
TYPE A[2]	3/4:1 (53 Deg.)
TYPE B	1:1 (45 Deg.)
TYPE C	1 1/2:1 (34 Deg.)

[1] Numbers shown in parentheses next to maximum allowable slopes are angles expressed in degrees from the horizontal. Angles have been rounded off.
[2] A short-term maximum allowable slope of 1/2H:1V (63 degrees)is allowed in excavations in Type A soil that are 12 feed (3.67 m) or less in depth. Short-term maximum allowable slopes for excavations greater than 12 feet (3.67 m) in depth shall be 3/4H:1V (53 degrees).
[3] Sloping or benching for excavations greater than 20 feet deep shall be designed by a registered professional engineer.

Figure B-1 - Slope Configurations

(All slopes stated below are in the horizontal to vertical ratio)

B-1.1 Excavations made in Type A soil.

1. All simple slope excavation 20 feet or less in depth shall have a maximum allowable slope of 3/4:1.

SIMPLE SLOPE – GENERAL

(Figure B-1.1a)

Exception: Simple slope excavations which are open 24 hours or less (short term) and which are 12 feet or less in depth shall have a maximum allowable slope of 1/2:1.

SIMPLE SLOPE - SHORT TERM

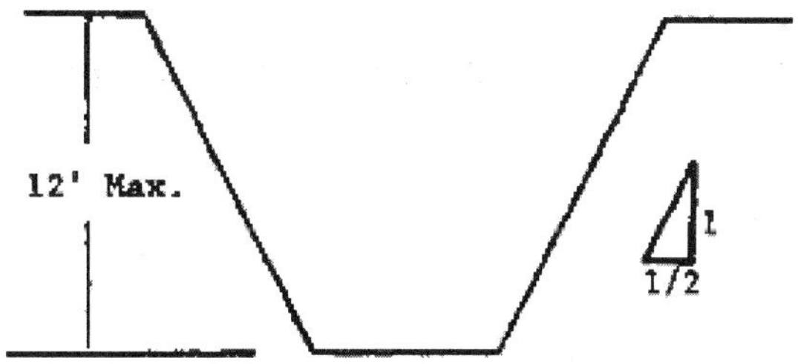

(Figure B-1.1b)

2. All benched excavations 20 feet or less in depth shall have a maximum allowable slope of 3/4 to 1 and maximum bench dimensions as follows:

SIMPLE BENCH

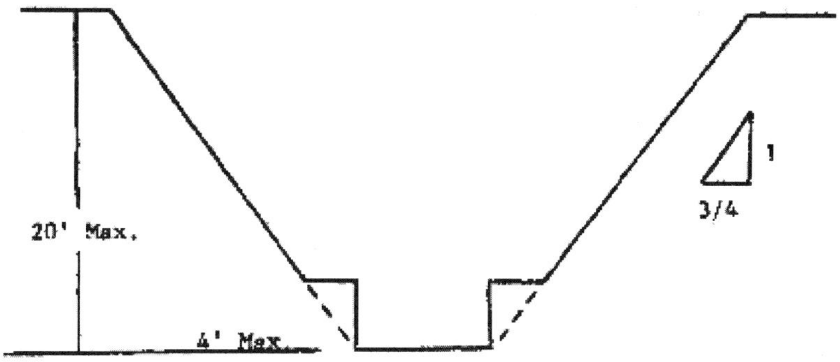

(Figure B-1.1c)

MULTIPLE BENCH

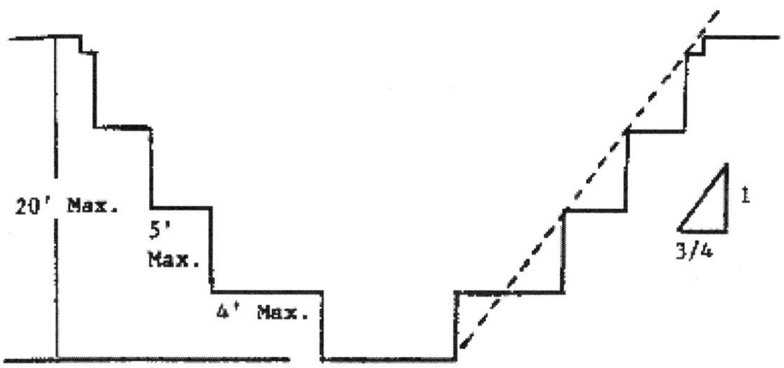

(Figure B-1.1d)

3. All excavations 8 feet or less in depth which have unsupported vertically sided lower portions shall have a maximum vertical side of 3 1/2 feet.

UNSUPPORTED VERTICALLY SIDED LOWER PORTION – (MAXIMUM 8 FEET IN DEPTH)

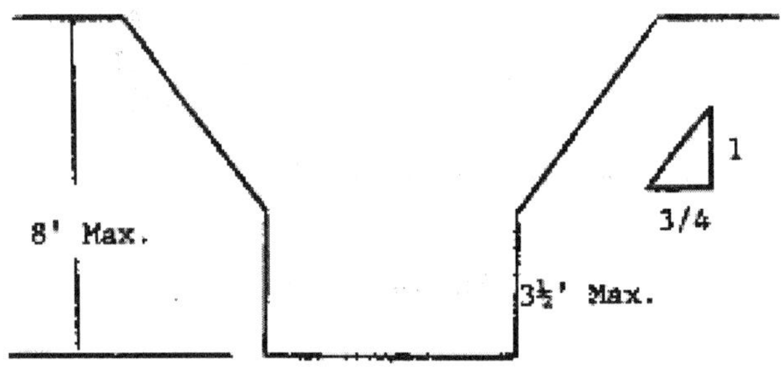

(Figure B-1.1e)

All excavations more than 8 feet but not more than 12 feet in depth with unsupported vertically sided lower portions shall have a maximum allowable slope of 1:1 and a maximum vertical side of 3 1/2 feet.

UNSUPPORTED VERTICALLY SIDED LOWER PORTION–(MAXIMUM 12 FEET IN DEPTH)

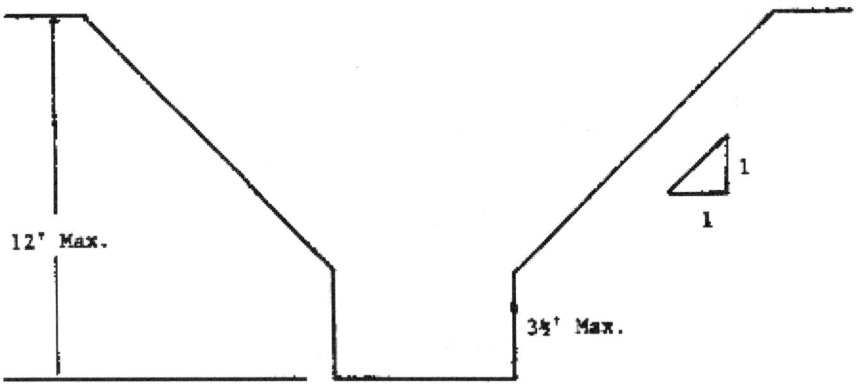

(Figure B-1.1f)

All excavations 20 feet or less in depth which have vertically sided lower portions that are supported or shielded shall have a maximum allowable slope of 3/4:1. The support or shield system must extend at least 18 inches above the top of the vertical side.

SUPPORTED OR SHIELDED VERTICALLY SIDED LOWER PORTION

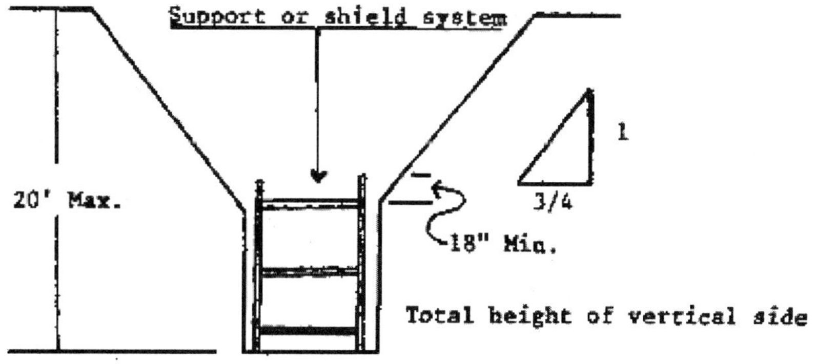

(Figure B-1.1g)

4. All other simple slope, compound slope, and vertically sided lower portion excavations shall be in accordance with the other options permitted under 1926.652(b).

B-1.2 Excavations Made in Type B Soil

1. All simple slope excavations 20 feet or less in depth shall have a maximum allowable slope of 1:1.

SIMPLE SLOPE

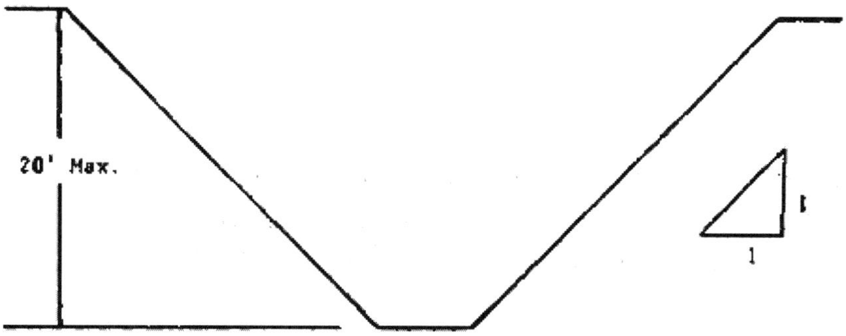

(Figure B-1.2a)

2. All benched excavations 20 feet or less in depth shall have a maximum allowable slope of 1:1 and maximum bench dimensions as follows:

SINGLE BENCH

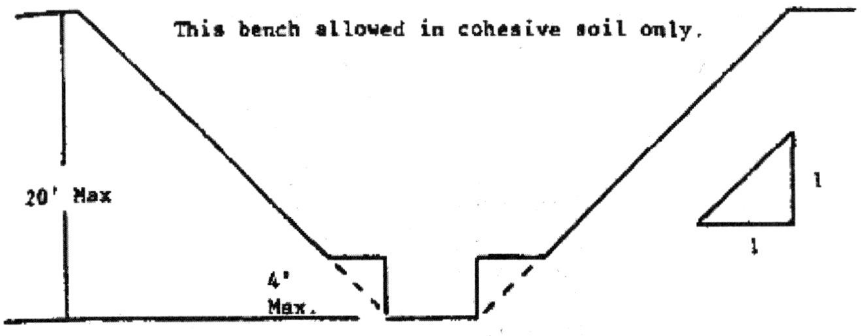

(Figure B-1.2b)

MULTIPLE BENCH

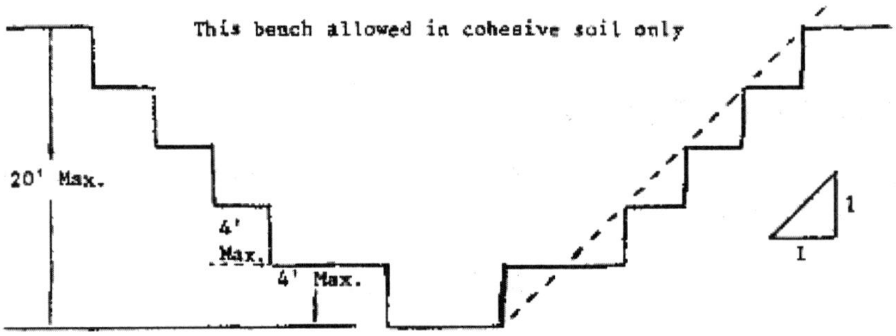

(Figure B-1.2c)

3. All excavations 20 feet or less in depth which have vertically sided lower portions shall be shielded or supported to a height at least 18 inches above the top of the vertical side. All such excavations shall have a maximum allowable slope of 1:1.

VERTICALLY SIDED LOWER PORTION

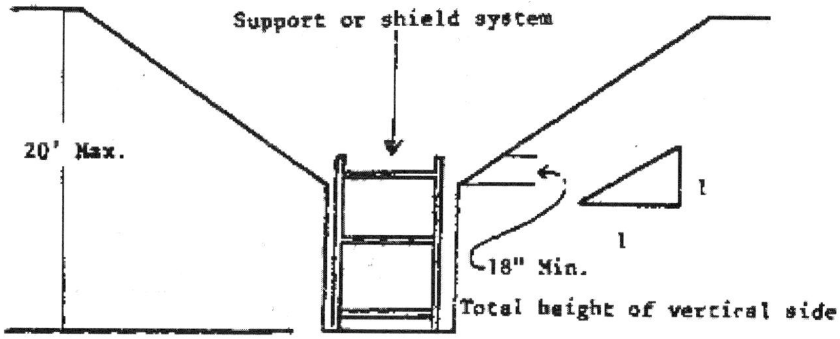

(Figure B-1.2d)

4. All other sloped excavations shall be in accordance with the other options permitted in 1926.652(b).

B - 1.3 Excavations Made in Type C Soil

1. All simple slope excavations 20 feet or less in depth shall have a maximum allowable slope of 1 1/2:1.

SIMPLE SLOPE

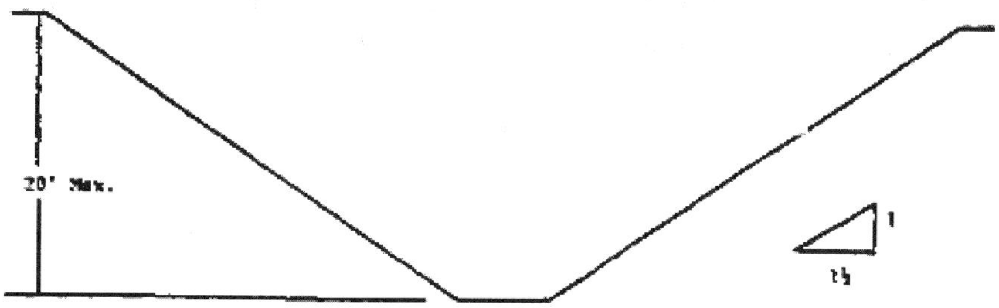

(Figure B-1.3a)

2. All excavations 20 feet or less in depth which have vertically sided lower portions shall be shielded or supported to a height at least 18 inches above the top of the vertical side. All such excavations shall have a maximum allowable slope of 1 1/2:1.

VERTICAL SIDED LOWER PORTION

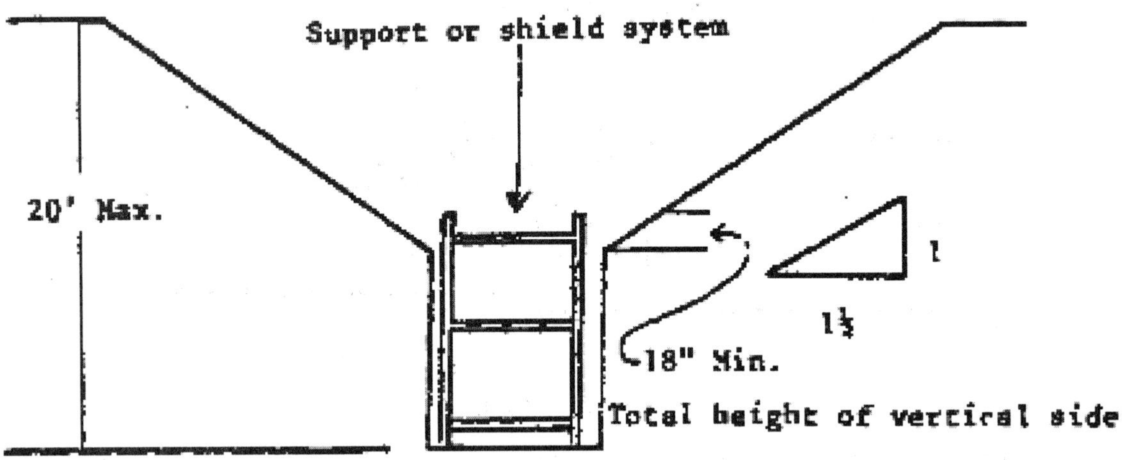

(Figure B-1.3b)

3. All other sloped excavations shall be in accordance with the other options permitted in 1926.652(b).

B - 1.4 Excavations Made in Layered Soils

1. All excavations 20 feet or less in depth made in layered soils shall have a maximum allowable slope for each layer as set forth below.

B OVER A

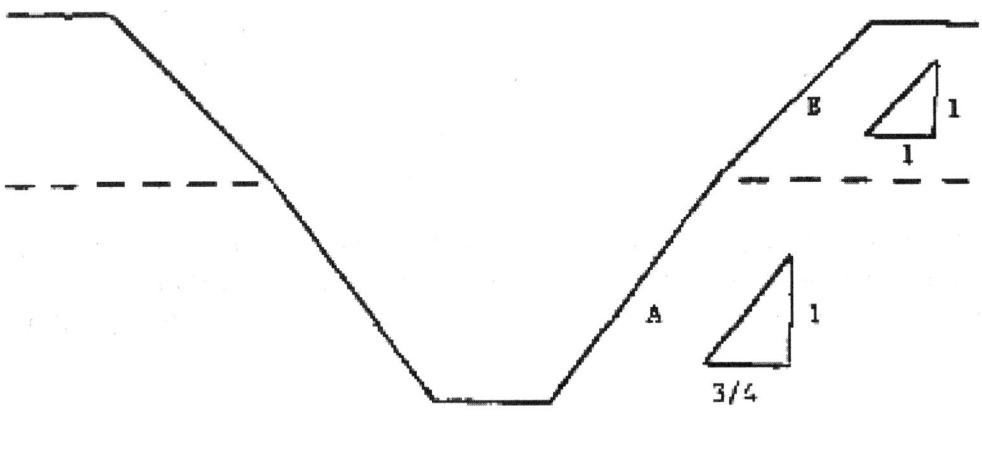

(Figure B-1.4a)

C OVER A

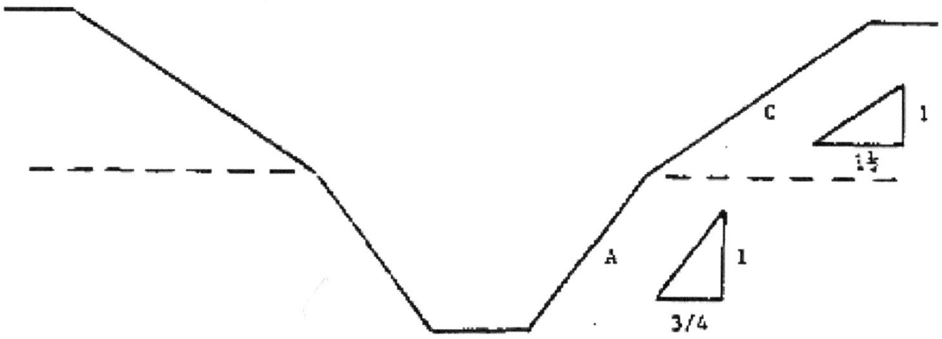

(Figure B-1.4b)

C OVER B

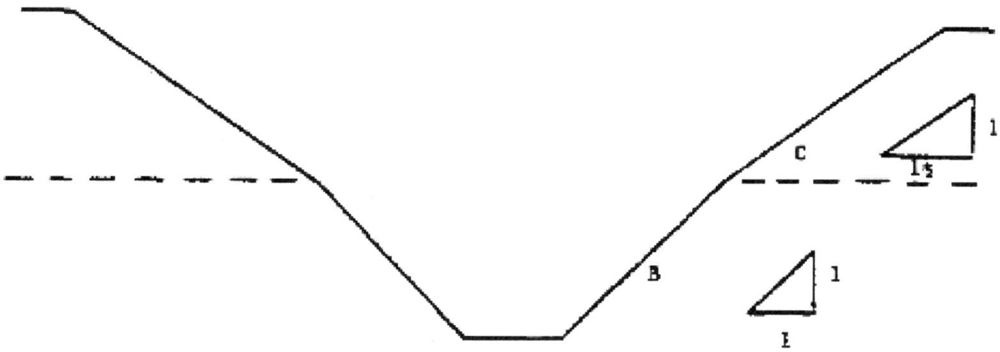

(Figure B-1.4c)

A OVER B

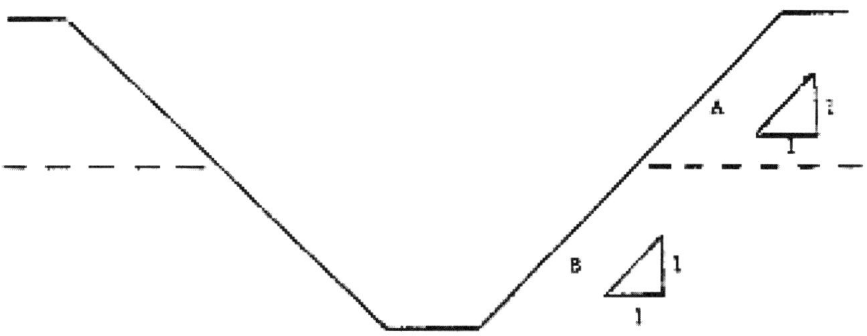

(Figure B-1.4d)

A OVER C

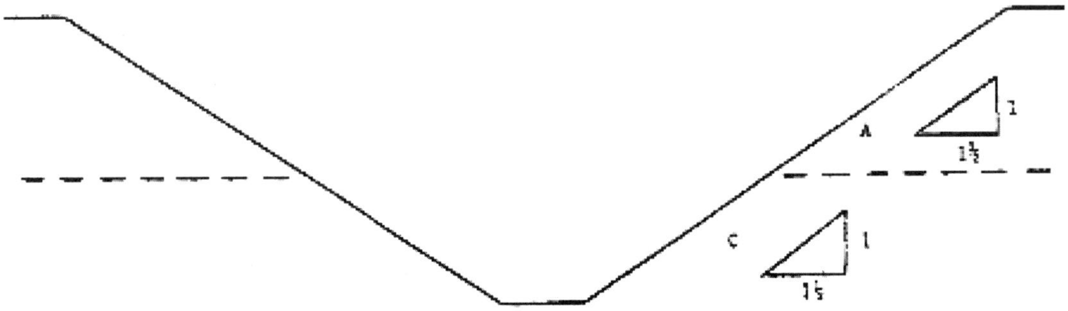

(Figure B-1.4e)

B OVER C

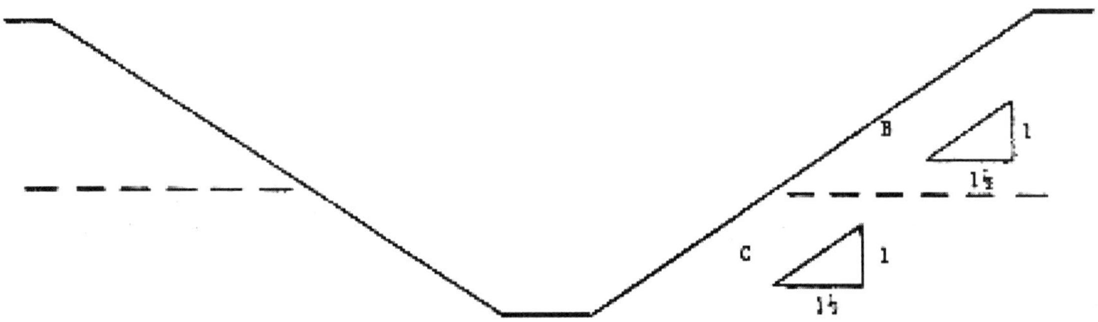

(Figure B-1.4f)

2. All other sloped excavations shall be in accordance with the other options permitted in 1926.652(b).

4.F. – 29 CFR 1926 Subpart P Appendix C – Timber Shoring for Trenches

• Part Number:	1926
• Part Title:	Safety and Health Regulations for Construction
• Subpart:	P
• Subpart Title:	Excavations
• Standard Number:	1926 Subpart P App C
• Title:	Timber Shoring for Trenches

(a) Scope.

This appendix contains information that can be used when timber shoring is provided as a method of protection from cave-ins in trenches that do not exceed 20 feet (6.1 m) in depth. This appendix must be used when design of timber shoring protective systems is to be performed in accordance with 1926.652(c)(1). Other timber shoring configurations; other systems of support such as hydraulic and pneumatic systems; and other protective systems such as sloping, benching, shielding, and freezing systems must be designed in accordance with the requirements set forth in 1926.652(b) and 1926.652(c).

(b) Soil Classification.

In order to use the data presented in this appendix, the soil type or types in which the excavation is made must first be determined using the soil classification method set forth in appendix A of subpart P of this part.

(c) Presentation of Information.

Information is presented in several forms as follows:

(1) Information is presented in tabular form in Tables C-1.1, C-1.2 and C-1.3, and Tables C-2.1, C-2.2 and C-2.3 following paragraph (g) of the appendix. Each table presents the minimum sizes of timber members to use in a shoring system, and each table contains data only for the particular soil type in which the excavation or portion of the excavation is made. The data are arranged to allow the user the flexibility to select from among several acceptable configurations of members based on varying the horizontal spacing of the crossbraces. Stable rock is exempt from shoring requirements and therefore, no data are presented for this condition.

(2) Information concerning the basis of the tabular data and the limitations of the data is presented in paragraph (d) of this appendix, and on the tables themselves.

(3) Information explaining the use of the tabular data is presented in paragraph (e) of this appendix.

(4) Information illustrating the use of the tabular data is presented in paragraph (f) of this appendix.

(5) Miscellaneous notations regarding Tables C-1.1 through C-1.3 and Tables C-2.1 through C-2.3 are presented in paragraph (g) of this Appendix.

(d) Basis and Limitations of the Data.

(1) Dimensions of timber members.

(i) The sizes of the timber members listed in Tables C-1.1 through C-1.3 are taken from the National Bureau of Standards (NBS) report, "Recommended Technical Provisions for Construction Practice in Shoring and Sloping of Trenches and Excavations." In addition, where NBS did not recommend specific sizes of members, member sizes are based on an analysis of the sizes required for use by existing codes and on empirical practice.

(ii) The required dimensions of the members listed in Tables C-1.1 through C-1.3 refer to actual dimensions and not nominal dimensions of the timber. Employers wanting to use nominal size shoring are directed to Tables C-2.1 through C-2.3, or have this choice under 1926.652(c)(3), and are referred to The Corps of engineers, The Bureau of Reclamation or data from other acceptable sources.

(2) Limitation of application.

(i) It is not intended that the timber shoring specification apply to every situation that may be experienced in the field. These data were developed to apply to the situations that are most commonly experienced in current trenching practice. Shoring systems for use in situations that are not covered by the data in this appendix must be designed as specified in 1926.652(c).

(ii) When any of the following conditions are present, the members specified in the tables are not considered adequate. Either an alternate timber shoring system must be designed or another type of protective system designed in accordance with 1926.652.

(A) When loads imposed by structures or by stored material adjacent to the trench weigh in excess of the load imposed by a two-foot soil surcharge. The term "adjacent" as used here means the area within a horizontal distance from the edge of the trench equal to the depth of the trench.

(B) When vertical loads imposed on cross braces exceed a 240-pound gravity load distributed on a one-foot section of the center of the crossbrace.

(C) When surcharge loads are present from equipment weighing in excess of 20,000 pounds.

(D) When only the lower portion of a trench is shored and the remaining portion of the trench is sloped or benched unless: The sloped portion is sloped at an angle less steep than three horizontal to one vertical; or the members are selected from the tables for use at a depth which is determined from the top of the overall trench, and not from the toe of the sloped portion.

(e) Use of Tables.

The members of the shoring system that are to be selected using this information are the cross braces, the uprights, and the wales, where wales are required. Minimum sizes of members are specified for use in different types of soil. There are six tables of information, two for each soil type. The soil type must first be determined in accordance with the soil classification system described in appendix A to subpart P of part 1926. Using the appropriate table, the selection of the size and spacing of the members is then made. The selection is based on the depth and width of the trench where the members are to be installed and, in most instances, the selection is also based on the horizontal spacing of the crossbraces. Instances where a choice of horizontal spacing of crossbracing is available, the horizontal spacing of the crossbraces must be chosen by the user before the size of any member can be determined. When the soil type, the width and depth of the trench, and the horizontal spacing of the crossbraces are known, the size and vertical spacing of the crossbraces are known, the size and vertical spacing of the crossbraces, the size and vertical spacing of the wales, and the size and horizontal spacing of the uprights can be read from the appropriate table.

(f) Examples to Illustrate the Use of Tables C-1.1 through C-1.3.

(1) Example 1.

A trench dug in Type A soil is 13 feet deep and five feet wide. From Table C-1.1, four acceptable arrangements of timber can be used.

Arrangement #1

Space 4x4 crossbraces at six feet horizontally and four feet vertically.

Wales are not required.

Space 3X8 uprights at six feet horizontally. This arrangement is commonly called "skip shoring."

Arrangement #2

Space 4x6 crossbraces at eight feet horizontally and four feet vertically.

Space 8x8 wales at four feet vertically.

Space 2x6 uprights at four feet horizontally.

Arrangement #3

Space 6x6 crossbraces at 10 feet horizontally and four feet vertically.

Space 8x10 wales at four feet vertically.

Space 2x6 uprights at five feet horizontally.

Arrangement #4

Space 6x6 crossbraces at 12 feet horizontally and four feet vertically.

Space 10x10 wales at four feet vertically.

Space 3x8 uprights at six feet horizontally.

(2) Example 2.

A trench dug in Type B soil is 13 feet deep and five feet wide. From Table C-1.2 three acceptable arrangements of members are listed.

Arrangement #1

Space 6X6 crossbraces at six feet horizontally and five feet vertically.

Space 8X8 wales at five feet vertically.

Space 2X6 uprights at two feet horizontally.

Arrangement #2

Space 6X8 crossbraces at eight feet horizontally and five feet vertically.

Space 10X10 wales at five feet vertically.

Space 2X6 uprights at two feet horizontally.

Arrangement #3

Space 8X8 crossbraces at 10 feet horizontally and five feet vertically.

Space 10X12 wales at five feet vertically.

Space 2X6 uprights at two feet vertically.

(3) Example 3.

A trench dug in Type C soil is 13 feet deep and five feet wide.

From Table C-1.3 two acceptable arrangements of members can be used.

Arrangement #1

Space 8X8 crossbraces at six feet horizontally and five feet vertically.

Space 10X12 wales at five feet vertically.

Position 2X6 uprights as closely together as possible.

If water must be retained use special tongue and groove uprights to form tight sheeting.

Arrangement #2

Space 8X10 crossbraces at eight feet horizontally and five feet vertically.

Space 12X12 wales at five feet vertically.

Position 2X6 uprights in a close sheeting configuration unless water pressure must be resisted. Tight sheeting must be used where water must be retained.

(4) Example 4.

A trench dug in Type C soil is 20 feet deep and 11 feet wide. The size and spacing of members for the section of trench that is over 15 feet in depth is determined using Table C-1.3. Only one arrangement of members is provided.

Space 8X10 crossbraces at six feet horizontally and five feet vertically.

Space 12X12 wales at five feet vertically.

Use 3X6 tight sheeting.

Use of Tables C-2.1 through C-2.3 would follow the same procedures.

(g) Notes for all Tables

1. Member sizes at spacings other than indicated are to be determined as specified in 1926.652(c), "Design of Protective Systems."
2. When conditions are saturated or submerged use Tight Sheeting. Tight Sheeting refers to the use of specially-edged timber planks (e.g., tongue and groove) at least three inches thick, steel sheet piling, or similar construction that when driven or placed in position provide a tight wall to resist the lateral pressure of water and to prevent the loss of backfill material. Close Sheeting refers to the placement of planks side-by-side allowing as little space as possible between them.
3. All spacing indicated is measured center to center.
4. Wales to be installed with greater dimension horizontal.
5. If the vertical distance from the center of the lowest crossbrace to the bottom of the trench exceeds two and one-half feet, uprights shall be firmly embedded or a mudsill shall be used. Where uprights are embedded, the vertical distance from the center of the lowest crossbrace to the bottom of the trench shall not exceed 36 inches. When mudsills are used, the vertical distance shall not exceed 42 inches. Mudsills are wales that are installed at the tow of the trench side.
6. Trench jacks may be used in lieu of or in combination with timber crossbraces.
7. Placement of crossbraces. When the vertical spacing of crossbraces is four feet, place the top crossbrace no more than two feet below the top of the trench. When the vertical spacing of crossbraces is five feet, place the top crossbrace no more than 2.5 feet below the top of the trench.

TABLE C-1.1. TIMBER TRENCH SHORING — MINIMUM TIMBER REQUIREMENTS *
SOIL TYPE A P(a) = 25 X H + 72 psf (2 ft Surcharge)

Depth of Trench (Feet)	SIZE (ACTUAL) AND SPACING OF MEMBERS **													
	CROSS BRACES							WALES		UPRIGHTS				
	Horiz. Spacing (Feet)	Width of Trench (Feet)					Vert. Spacing (Feet)	Size (in.)	Vert. Spacing (Feet)	Maximim Allowable Horizontal Spacing (Feet)				
		Up to 4	Up to 6	Up to 9	Up to 12	Up to 15				Close	4	5	6	8
5 to 10	Up to 6	4x4	4x4	4x6	6x6	6x6	4	Not req'd					2X6	
	Up to 8	4x4	4x4	4x6	6x6	6x6	4	Not req'd						2X8
	Up to 10	4x6	4x6	4x6	6x6	6x6	4	8x8	4			2X6		
	Up to 12	4x6	4x6	6x6	6x6	6x6	4	8x8	4				2X6	
10 to 15	Up to 6	4x4	4x4	4x6	6x6	6x6	4	Not req'd					3X8	
	Up to 8	4x6	4x6	6x6	6x6	6x6	4	8x8	4		2X6			
	Up to 10	6x6	6x6	6x6	6x8	6x8	4	8x10	4			2X6		
	Up to 12	6x6	6x6	6x6	6x8	6x8	4	10x10	4				3X8	
15 to 20	Up to 6	6x6	6x6	6x6	6x8	6x8	4	6x8	4	3X6				
	Up to 8	6x6	6x6	6x6	6x8	6x8	4	8x8	4	3X6				
	Up to 10	8x8	8x8	8x8	8x8	8x10	4	8x10	4	3X6				
	Up to 12	8x8	8x8	8x8	8x8	8x10	4	10x10	4	3X6				
Over 20	SEE NOTE 1													

* Mixed oak or equivalent with a bending strength not less than 850 psi.
** Manufactured members of equivalent strength may be substituted for wood.

TABLE C-1.2. TIMBER TRENCH SHORING — MINIMUM TIMBER REQUIREMENTS *
SOIL TYPE B P(a) = 45 X H + 72 psf (2 ft Surcharge)

Depth of Trench (Feet)	SIZE (ACTUAL) AND SPACING OF MEMBERS **													
	CROSS BRACES							WALES		UPRIGHTS				
	Horiz. Spacing (Feet)	Width of Trench (Feet)					Vert. Spacing (Feet)	Size (in.)	Vert. Spacing (Feet)	Maximim Allowable Horizontal Spacing (Feet)				
		Up to 4	Up to 6	Up to 9	Up to 12	Up to 15				Close	2	3		
5 to 10	Up to 6	4x6	4x6	6x6	6x6	6x6	5	6x8	5			2x6		
	Up to 8	6x6	6x6	6x6	6x8	6x8	5	8x10	5			2x6		
	Up to 10	6x6	6x6	6x6	6x8	6x8	5	10x10	5			2x6		
	see note 1													
10 to 15	Up to 6	6x6	6x6	6x6	6x8	6x8	5	8x8	5		2x6			
	Up to 8	6x8	6x8	6x8	8x8	8x8	5	10x10	5		2x6			
	Up to 10	8x8	8x8	8x8	8x8	8x10	5	10x12	5		2x6			
	see note 1													
15 to 20	Up to 6	6x8	6x8	6x8	8x8	8x8	5	8x10	5	3X6				
	Up to 8	8x8	8x8	8x8	8x8	8x10	5	10x12	5	3X6				
	Up to 10	8x10	8x10	8x10	8x10	10x10	5	12x12	5	3X6				
	see note 1													
Over 20	SEE NOTE 1													

* Mixed oak or equivalent with a bending strength not less than 850 psi.
** Manufactured members of equivalent strength may be substituted for wood.

TABLE C-1.3. TIMBER TRENCH SHORING — MINIMUM TIMBER REQUIREMENTS *
SOIL TYPE C P(a) = 80 X H + 72 psf (2 ft Surcharge)

Depth of Trench (Feet)	SIZE (ACTUAL) AND SPACING OF MEMBERS **													
	CROSS BRACES							WALES		UPRIGHTS				
	Horiz. Spacing (Feet)	Width of Trench (Feet)					Vert. Spacing (Feet)	Size (in.)	Vert. Spacing (Feet)	Maximim Allowable Horizontal Spacing (Feet)				
		Up to 4	Up to 6	Up to 9	Up to 12	Up to 15				Close	2	3		
5 to 10	Up to 6	6x8	6x8	6x8	8x8	8x8	5	8x10	5	2x6				
	Up to 8	8x8	8x8	8x8	8x8	8x10	5	10x12	5	2x6				
	Up to 10	8x10	8x10	8x10	8x10	10x10	5	12x12	5	2x6				
	see note 1													
10 to 15	Up to 6	8x8	8x8	8x8	8x8	8x10	5	10x10	5	2x6				
	Up to 8	8x10	8x10	8x10	8x10	10x10	5	10x12	5	2x6				
	see note 1													
	see note 1													
15 to 20	Up to 6	8x10	8x10	8x10	8x10	10x10	5	12x12	5	3x6				
	see note 1													
	see note 1													
	see note 1													
Over 20	SEE NOTE 1													

* Mixed oak or equivalent with a bending strength not less than 850 psi.

** Manufactured members of equivalent strength may be substituted for wood.

TABLE C-2.1. TIMBER TRENCH SHORING — MINIMUM TIMBER REQUIREMENTS *
SOIL TYPE A P(a) = 25 X H + 72 psf (2 ft Surcharge)

Depth of Trench (Feet)	SIZE (S4S) AND SPACING OF MEMBERS **													
	CROSS BRACES							WALES		UPRIGHTS				
	Horiz. Spacing (Feet)	Width of Trench (Feet)					Vert. Spacing (Feet)	Size (in.)	Vert. Spacing (Feet)	Maximim Allowable Horizontal Spacing (Feet)				
		Up to 4	Up to 6	Up to 9	Up to 12	Up to 15				Close	4	5	6	8
5 to 10	Up to 6	4x4	4x4	4x4	4x4	4x6	4	Not req'd	Not req'd				4x6	
	Up to 8	4x4	4x4	4x4	4x6	4x6	4	Not req'd	Not req'd					4x8
	Up to 10	4x6	4x6	4x6	6x6	6x6	4	8x8	4			4x6		
	Up to 12	4x6	4x6	4x6	6x6	6x6	4	8x8	4				4x6	
10 to 15	Up to 6	4x4	4x4	4x4	6x6	6x6	4	Not req'd	Not req'd				4x10	
	Up to 8	4x6	4x6	4x6	6x6	6x6	4	6x8	4		4x6			
	Up to 10	6x6	6x6	6x6	6x6	6x6	4	8x8	4			4x8		
	Up to 12	6x6	6x6	6x6	6x6	6x6	4	8x10	4		4x6		4x10	
15 to 20	Up to 6	6x6	6x6	6x6	6x6	6x6	4	6x8	4	3x6				
	Up to 8	6x6	6x6	6x6	6x6	6x6	4	8x8	4	3x6				
	Up to 10	6x6	6x6	6x6	6x6	6x8	4	8x10	4	3x6				
	Up to 12	6x6	6x6	6x6	6x8	6x8	4	8x12	4	3x6				
Over 20	SEE NOTE 1													

* Douglas fir or equivalent with a bending strength not less than 1500 psi.
** Manufactured members of equivalent strength may be substituted for wood.

TABLE C-2.2. TIMBER TRENCH SHORING — MINIMUM TIMBER REQUIREMENTS *
SOIL TYPE B P(a) = 45 X H + 72 psf (2 ft Surcharge)

Depth of Trench (Feet)	SIZE (S4S) AND SPACING OF MEMBERS **													
	CROSS BRACES							WALES		UPRIGHTS				
	Horiz. Spacing (Feet)	Width of Trench (Feet)					Vert. Spacing (Feet)	Size (in.)	Vert. Spacing (Feet)	Maximim Allowable Horizontal Spacing (Feet)				
		Up to 4	Up to 6	Up to 9	Up to 12	Up to 15				Close	2	3		
5 to 10	Up to 6	4x6	4x6	4x6	6x6	6x6	5	6x8	5			3x12 4x8		4x12
	Up to 8	4x6	4x6	6x6	6x6	6x6	5	8x8	5		3x8		4x8	
	Up to 10	4x6	4x6	6x6	6x6	6x8	5	8x10	5			4x8		
	see note 1													
10 to 15	Up to 6	6x6	6x6	6x6	6x8	6x8	5	8x8	5	3x6	4x10			
	Up to 8	6x8	6x8	6x8	8x8	8x8	5	10x10	5	3x6	4x10			
	Up to 10	6x8	6x8	8x8	8x8	8x8	5	10x12	5	3x6	4x10			
	see note 1													
15 to 20	Up to 6	6x8	6x8	6x8	6x8	8x8	5	8x10	5	4x6				
	Up to 8	6x8	6x8	6x8	8x8	8x8	5	10x12	5	4x6				
	Up to 10	8x8	8x8	8x8	8x8	8x8	5	12x12	5	4x6				
	see note 1													
Over 20	SEE NOTE 1													

* Douglas fir or equivalent with a bending strength not less than 1500 psi.
** Manufactured members of equivalent strength may be substituted for wood.

TABLE C-2.3. TIMBER TRENCH SHORING — MINIMUM TIMBER REQUIREMENTS *
SOIL TYPE C P(a) = 80 X H + 72 psf (2 ft Surcharge)

Depth of Trench (Feet)	SIZE (S4S) AND SPACING OF MEMBERS **														
	CROSS BRACES							WALES		UPRIGHTS				
	Horiz. Spacing (Feet)	Width of Trench (Feet)					Vert. Spacing (Feet)	Size (in.)	Vert. Spacing (Feet)	Maximim Allowable Horizontal Spacing (Feet)				
		Up to 4	Up to 6	Up to 9	Up to 12	Up to 15				Close				
5 to 10	Up to 6	6x6	6x6	6x6	6x6	8x8	5	8x8	5	3x6				
	Up to 8	6x6	6x6	6x6	8x8	8x8	5	10x10	5	3x6				
	Up to 10	6x6	6x6	8x8	8x8	8x8	5	10x12	5	3x6				
	see note 1													
10 to 15	Up to 6	6x8	6x8	6x8	8x8	8x8	5	10x10	5	4x6				
	Up to 8	8x8	8x8	8x8	8x8	8x8	5	12x12	5	4x6				
	see note 1													
	see note 1													
15 to 20	Up to 6	8x8	8x8	8x8	8x10	8x10	5	10x12	5	4x6				
	see note 1													
	see note 1													
	see note 1													
Over 20	SEE NOTE 1													

* Douglas fir or equivalent with a bending strength not less than 1500 psi.
** Manufactured members of equivalent strength may be substituted for wood.

4.G. – 29 CFR 1926 Subpart P Appendix D – Aluminum Hydraulic Shoring for Trenches

- Part Number: 1926
- Part Title: Safety and Health Regulations for Construction
- Subpart : P
- Subpart Title: Excavations
- Standard Number: 1926.650
- Title: Aluminum Hydraulic Shoring for Trenches

(a) Scope.

This appendix contains information that can be used when aluminum hydraulic shoring is provided as a method of protection against cave-ins in trenches that do not exceed 20 feet (6.1m) in depth. This appendix must be used when design of the aluminum hydraulic protective system cannot be performed in accordance with 1926.652(c)(2).

(b) Soil Classification.

In order to use data presented in this appendix, the soil type or types in which the excavation is made must first be determined using the soil classification method set forth in appendix A of subpart P of part 1926.

(c) Presentation of Information.

Information is presented in several forms as follows:

(1) Information is presented in tabular form in Tables D-1.1, D-1.2, D-1.3 and D-1.4. Each table presents the maximum vertical and horizontal spacings that may be used with various aluminum member sizes and various hydraulic cylinder sizes. Each table contains data only for the particular soil type in which the excavation or portion of the excavation is made. Tables D-1.1 and D-1.2 are for vertical shores in Types A and B soil. Tables D-1.3 and D-1.4 are for horizontal waler systems in Types B and C soil.

(2) Information concerning the basis of the tabular data and the limitations of the data is presented in paragraph (d) of this appendix.

(3) Information explaining the use of the tabular data is presented in paragraph (e) of this appendix.

(4) Information illustrating the use of the tabular data is presented in paragraph (f) of this appendix.

(5) Miscellaneous notations (Footnotes) regarding Table D-1.1 through D-1.4 are presented in paragraph (g) of this appendix.

(6) Figures, illustrating typical installations of hydraulic shoring, are included just prior to the Tables. The illustrations page is entitled "Aluminum Hydraulic Shoring: Typical Installations."

(d) Basis and Limitations of the Data.

(1) Vertical shore rails and horizontal wales are those that meet the Section Modulus requirements in the D-1 Tables. Aluminum material is 6061-T6 or material of equivalent strength and properties.

(2) Hydraulic cylinders specifications.

(i) 2-inch cylinders shall be a minimum 2-inch inside diameter with a minimum safe working capacity of no less than 18,000 pounds axial compressive load at maximum extension. Maximum extension is to include full range of cylinder extensions as recommended by product manufacturer.

(ii) 3-inch cylinders shall be a minimum 3-inch inside diameter with a safe working capacity of not less than 30,000 pounds axial compressive load at extensions as recommended by product manufacturer.

(3) Limitation of application.

(i) It is not intended that the aluminum hydraulic specification apply to every situation that may be experienced in the field. These data were developed to apply to the situations that are most commonly experienced in current trenching practice. Shoring systems for use in situations that are not covered by the data in this appendix must be otherwise designed as specified in 1926.652(c).

(ii) When any of the following conditions are present, the members specified in the Tables are not considered adequate. In this case, an alternative aluminum hydraulic shoring system or other type of protective system must be designed in accordance with 1926.652.

(A) When vertical loads imposed on cross braces exceed a 100 Pound gravity load distributed on a one foot section of the center of the hydraulic cylinder.

(B) When surcharge loads are present from equipment weighing in excess of 20,000 pounds.

(C) When only the lower portion of a trench is shored and the remaining portion of the trench is sloped or benched unless: The sloped portion is sloped at an angle less steep than three horizontal to one vertical; or the members are selected from the tables for use at a depth which is determined from the top of the overall trench, and not from the toe of the sloped portion.

(e) Use of Tables D-1.1, D-1.2, D-1.3 and D-1.4.

The members of the shoring system that are to be selected using this information are the hydraulic cylinders, and either the vertical shores or the horizontal wales. When a waler system is used the vertical timber sheeting to be used is also selected from these tables. The Tables D-1.1 and D-1.2 for vertical shores are used in Type A and B soils that do not require sheeting. Type B soils that may require sheeting, and Type C soils that always require sheeting, are found in the horizontal wale Tables D-1.3 and D-1.4. The soil type must first be determined in accordance with the soil classification system described in appendix A to subpart P of part 1926. Using the appropriate table, the selection of the size and spacing of the members is made. The selection is based on the depth and width of the trench where the members are to be installed. In these tables the vertical spacing is held constant at four feet on center. The tables show the maximum horizontal spacing of cylinders allowed for each size of wale in the waler system tables, and in the vertical shore tables, the hydraulic cylinder horizontal spacing is the same as the vertical shore spacing.

In regard to clarification of the proper spacing and placement of vertical hydraulic shoring members, prudent engineering practice would generally dictate that not more than one half of the allowable maximum horizontal spacing between vertical shores would be acceptable as the spacing between the end of the trench and the nearest vertical shore. Local soil structure and composition, however, determine the actual spacing and, subsequently, the required number of shores for a given length of trench, the degree of sloping of the ends, and any other safeguard that may be necessary to protect the worker. The footnotes and general notes to the appendix are minimum requirements and should not be interpreted as a listing of all the safeguards or information needed to protect employees working in trenches.

In regard to questions concerning the bending strength of plywood, the plywood size requirement listed in Appendix D, paragraph (g)(7) was derived to adequately control sloughing of the trench face in certain Type B soils. Any correlation between the bending strength of the plywood sheeting used in aluminum hydraulic shoring systems and the bending strength of the structural members of other shoring methods is coincidental.

It is permissible to use non-structural sheeting when it is not subject to the requirements of the regulations. However, the presence of such sheeting is likely to be interpreted as an indication that Type A or Type B soil that does not require sheeting are, in fact, not present. The sheeting is likely to be seen as an indication that other Type B or Type C soils are present and an inferior sheeting system is in place. Consequently, the contractor should maintain appropriate soil classification test results to minimize such misinterpretations.

(f) Example to Illustrate the Use of the Tables:

(1) Example 1:

A trench dug in Type A soil is 6 feet deep and 3 feet wide. From Table D-1.1: Find vertical shores and 2 inch diameter cylinders spaced 8 feet on center (o.c.) horizontally and 4 feet on center (o.c.) vertically. (See Figures 1 & 3 for typical installations.)

(2) Example 2:

A trench is dug in Type B soil that does not require sheeting, 13 feet deep and 5 feet wide.

From Table D-1.2: Find vertical shores and 2 inch diameter cylinders spaced 6.5 feet o.c. horizontally and 4 feet o.c. vertically. (See Figures 1 & 3 for typical installations.)

(3) A trench is dug in Type B soil that does not require sheeting, but does experience some minor raveling of the trench face. The trench is 16 feet deep and 9 feet wide. From Table D-1.2: Find vertical shores and 2 inch diameter cylinder (with special oversleeves as designated by Footnote #2) spaced 5.5 feet o.c. horizontally and 4 feet o.c. vertically. Plywood (per Footnote (g)(7) to the D-1 Table) should be used behind the shores. (See Figures 2 & 3 for typical installations.)

(4) Example 4: A trench is dug in previously disturbed Type B soil, with characteristics of a Type C soil, and will require sheeting. The trench is 18 feet deep, and 12 feet wide 8 foot horizontal spacing between cylinders is desired for working space. From Table D-1.3: Find horizontal wale with a section modulus of 14.0 spaced at 4 feet o.c. vertically and 3 inch diameter cylinder spaced at 9 feet maximum o.c. horizontally, 3 x 12 timber sheeting is required at close spacing vertically. (See Figure 4 for typical installation.)

(5) Example 5: A trench is dug in Type C soil, 9 feet deep and 4 feet wide. Horizontal cylinder spacing in excess of 6 feet is desired for working space. From Table D-1.4: Find horizontal wale with a section modulus of 7.0 and 2 inch diameter cylinders spaced at 6.5 feet o.c. horizontally. Or, find horizontal wale with a 14.0 section modulus and 3 inch diameter cylinder spaced at 10 feet o.c. horizontally. Both wales are spaced 4 feet o.c. vertically, 3 x 12 timber sheeting is required at close spacing vertically. (See Figure 4 for typical installation.)

(g) Footnotes, and general notes, for Tables D-1.1, D-1.2, D-1.3, and D-1.4.

(1) For applications other than those listed in the tables, refer to 1926.652(c)(2) for use of manufacturer's tabulated data. For trench depths in excess of 20 feet, refer to 1926.652(c)(2) and 1926.652(c)(3).

(2) 2 inch diameter cylinders, at this width, shall have structural steel tube (3.5 x 3.5 x 0.1875) oversleeves, or structural oversleeves of manufacturer's specification, extending the full, collapsed length.

(3) Hydraulic cylinders capacities.

(i) 2-inch cylinders shall be a minimum 2-inch inside diameter with a safe working capacity of not less than 18,000 pounds axial compressive load at maximum extension. Maximum extension is to include full range of cylinder extensions as recommended by product manufacturer.

(ii) 3-inch cylinders shall be a minimum 3-inch inside diameter with a safe work capacity of not less than 30,000 pounds axial compressive load at maximum extension. Maximum extension is to include full range of cylinder extensions as recommended by product manufacturer.

(4) All spacing indicated is measured center to center.

(5) Vertical shoring rails shall have a minimum section modulus of 0.40 inch.

(6) When vertical shores are used, there must be a minimum of three shores spaced equally, horizontally, in a group.

(7) Plywood shall be 1.125 inch thick softwood or 0.75 inch thick, 14 ply, arctic white birch (Finland form). Please note that plywood is not intended as a structural member, but only for prevention of local raveling (sloughing of the trench face) between shores.

As to the intent of note in 29 CFR, Part 1926, Subpart P, Appendix D, Aluminum Hydraulic Shoring for Trenches, please be advised that this note (g)(7) only applies to Tables D-1.1 through D-1.4 of this appendix where an employer chooses to design a protective system in accordance with 29 CFR 1926.652(c)(1), Option (1). The plywood mentioned in this note is not intended as a substitute for any primary load carrying structural members, e.g. timber uprights or aluminum rails as called for in Tables D-1.1 through D-1.4. In other words, the plywood in this note can only be used to contain raveled soil and to transfer the loads imposed by this soil to the load carrying members.

In regard to the use of equivalent strength plywood, substitutions of plywood thicknesses or types other than the two mentioned in note (g)(7) will be accepted for a system designed pursuant to 1926.652 (c)(1), Option (1), provided they have equal or better strength and stiffness than 1.125 inch softwood plywood or .75 inch 14-ply arctic white birch. The use of two 3/4 inch thick sheets of plywood with the spans parallel to the face grain is just one of the alternatives that might prove feasible. Any other plywood type or combination with equal or better strength and stiffness would satisfy the intent of the requirement listed in note (g)(7).

(8) See appendix C for timber specifications.

(9) Wales are calculated for simple span conditions.

(10) See appendix D, item (d), for basis and limitations of the data.

ALUMINUM HYDRAULIC SHORING
TYPICAL INSTALLATIONS

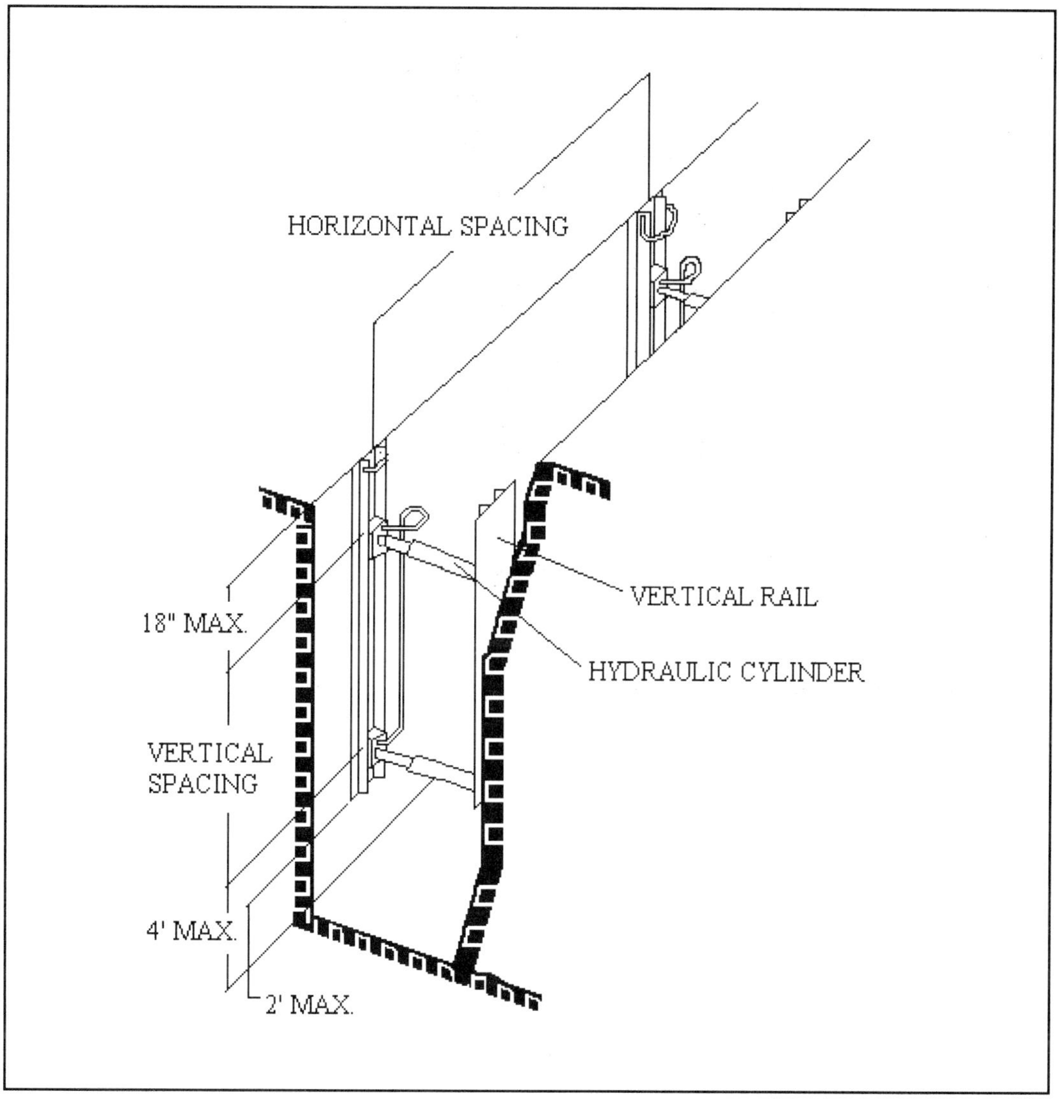

Figure No. 1 - Vertical Aluminum Hydraulic Shoring (Spot Bracing)

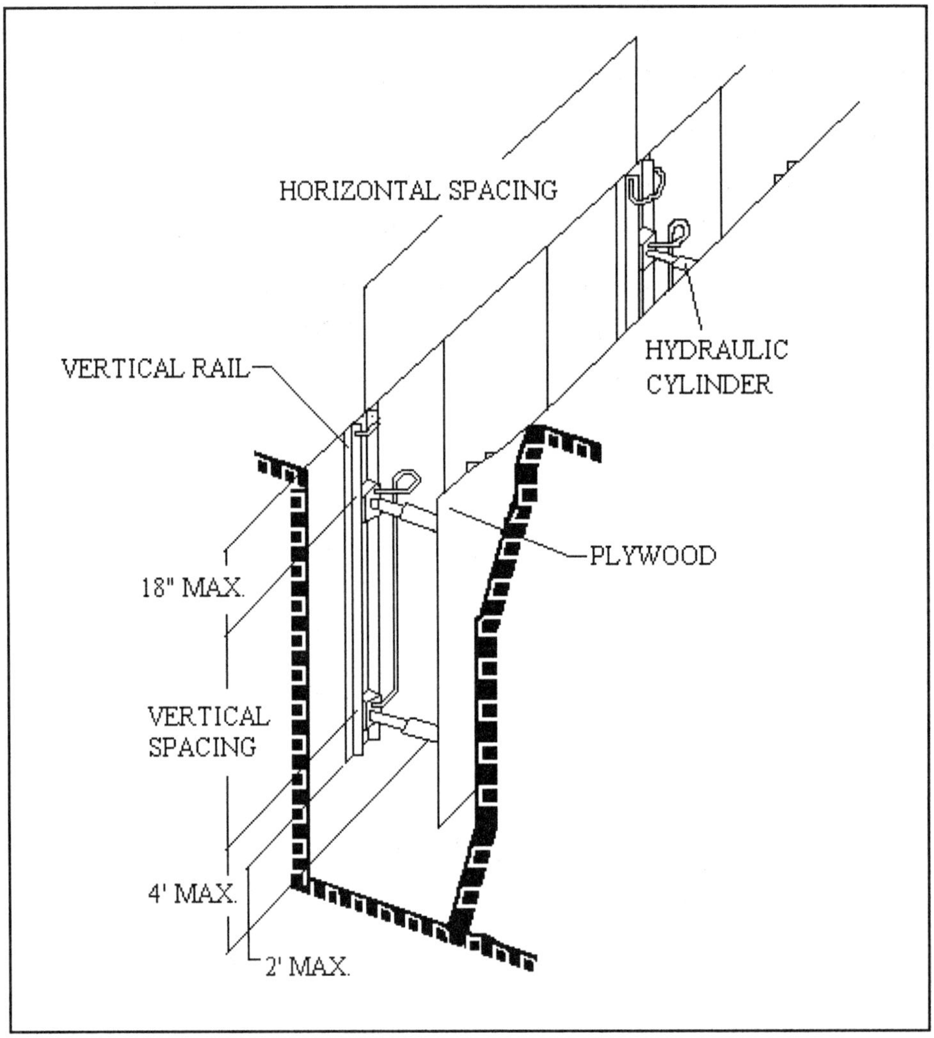

Figure No. 2 - Vertical Aluminum Hydraulic Shoring (with Plywood)

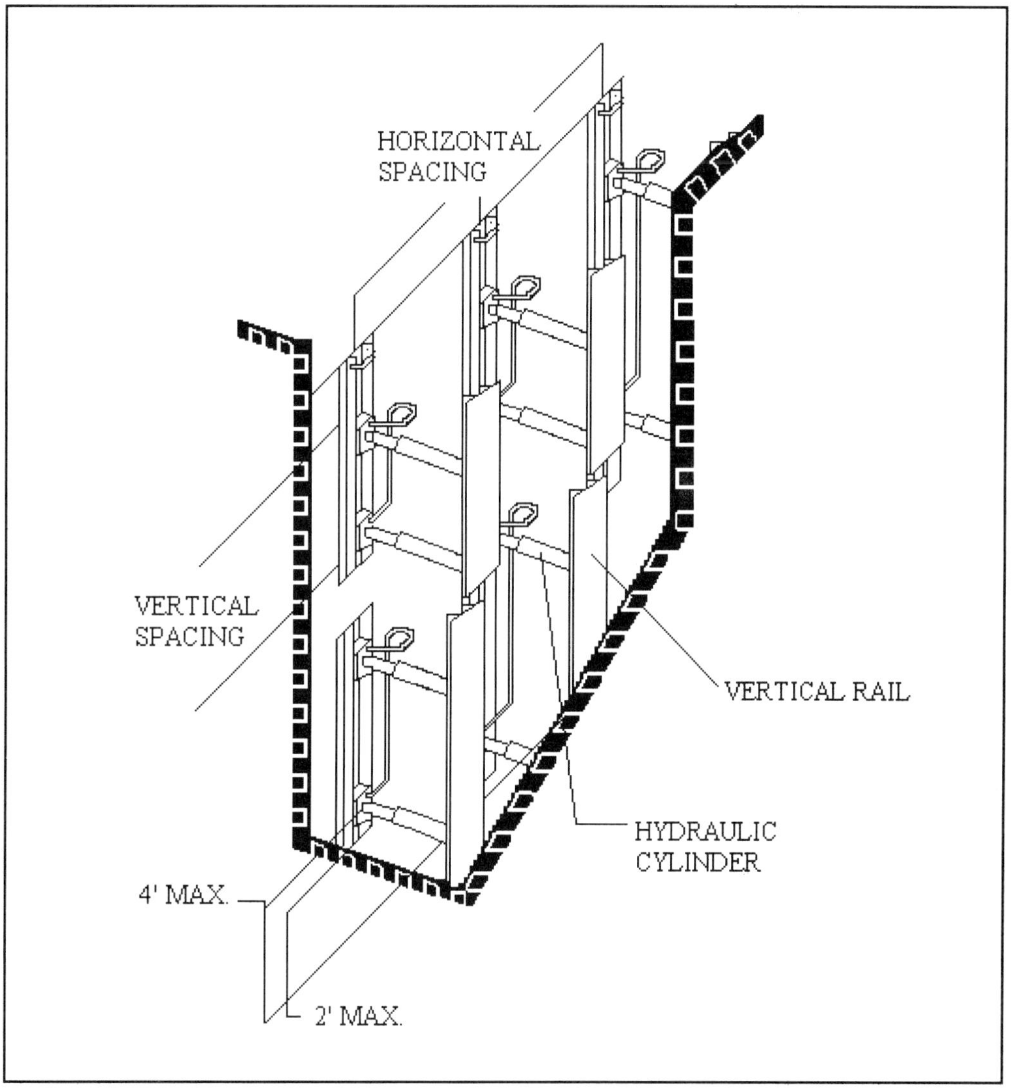

Figure No. 3 - Vertical Aluminum Hydraulic Shoring (Stacked)

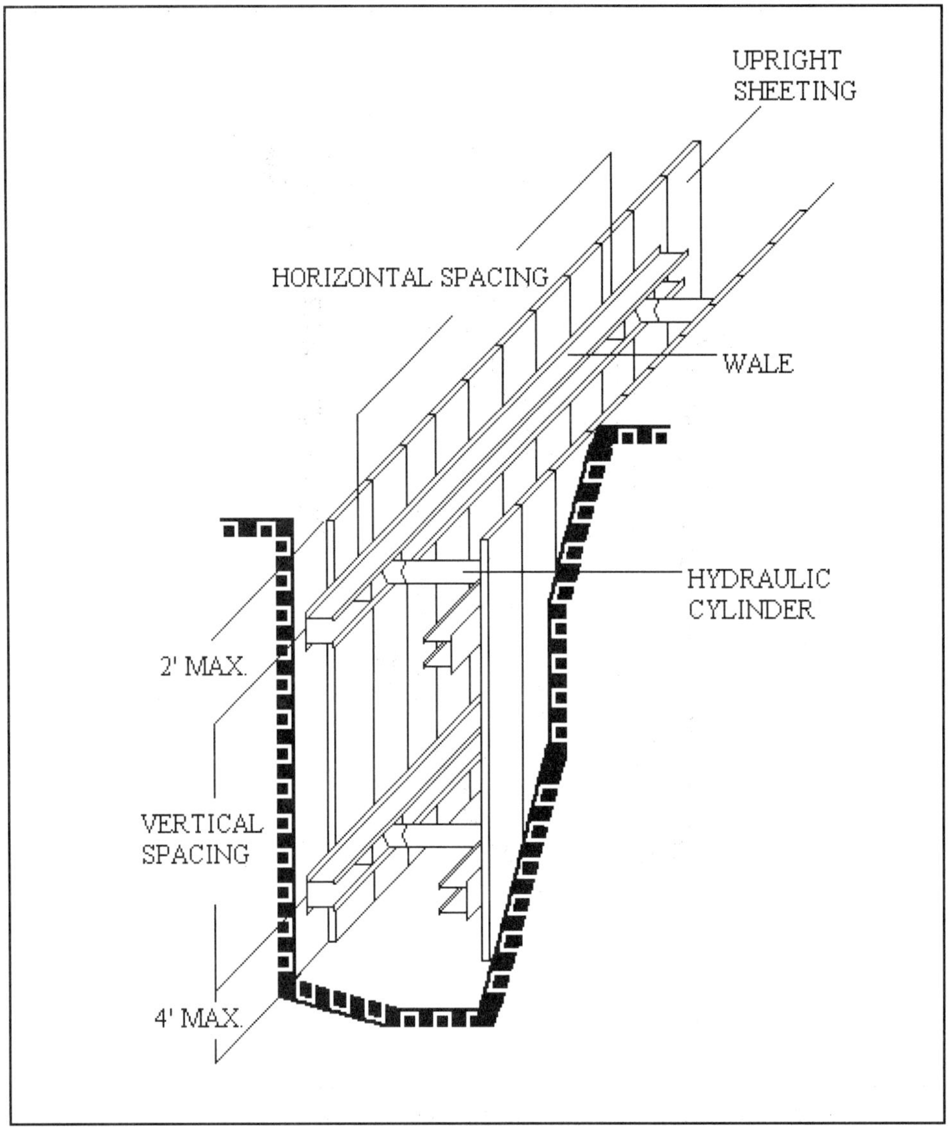

Figure No. 4 - Aluminum Hydraulic Shoring - Waler System (Typical)

TABLE D - 1.1. ALUMINUM HYDRAULIC SHORING VERTICAL SHORES FOR SOIL TYPE A

<table>
<tr><th rowspan="3">Depth of Trench (Feet)</th><th colspan="5">Hydraulic Cylinders</th></tr>
<tr><th rowspan="2">Maximum Horizontal Spacing (Feet)</th><th rowspan="2">Maximum Vertical Spacing (Feet)</th><th colspan="3">Width of Trench (feet)</th></tr>
<tr><th>Up to 8</th><th>Over 8 Up to 12</th><th>Over 12 Up to 15</th></tr>
<tr><td>Over 5 Up to 10</td><td>8</td><td rowspan="3">4</td><td rowspan="3">2 inch diameter</td><td rowspan="3">2 inch diameter Note (2)</td><td rowspan="3">3 inch diameter</td></tr>
<tr><td>Over 10 Up to 15</td><td>8</td></tr>
<tr><td>Over 15 Up to 20</td><td>7</td></tr>
<tr><td>Over 20</td><td colspan="5">See Note 1.</td></tr>
</table>

Footnotes to tables, and general notes on hydraulic shoring, are found in Appendix D, Item (g)

Note(1): See Appendix D, Item (g)(1)

Note(2): See Appendix D, Item (g)(2)

TABLE D - 1.2. ALUMINUM HYDRAULIC SHORING VERTICAL SHORES FOR SOIL TYPE B

Depth of Trench (Feet)	Hydraulic Cylinders				
	Maximum Horizontal Spacing (Feet)	Maximum Vertical Spacing (Feet)	Width of Trench (feet)		
			Up to 8	Over 8 Up to 12	Over 12 Up to 15
Over 5 Up to 10	8	4	2 inch diameter	2 inch diameter Note (2)	3 inch diameter
Over 10 Up to 15	6.5				
Over 15 Up to 20	5.5				
Over 20	See Note 1.				

Footnotes to tables, and general notes on hydraulic shoring, are found in Appendix D, Item (g)

Note(1): See Appendix D, Item (g)(1)

Note(2): See Appendix D, Item (g)(2)

TABLE D - 1.3. ALUMINUM HYDRAULIC SHORING WALER SYSTEMS FOR SOIL TYPE B

Depth of Trench (Feet)	WALES		HYDRAULIC CYLINDERS						TIMBER UPRIGHTS		
	Vertical Spacing (Feet)	Section Modulus (in(3))	Width of Trench (Feet)						Max. Horiz. Spacing (On Center)		
			Up to 8		Over 8 Up to 12		Over 12 Up to 15		Solid Sheet	2 ft	3 ft
			Horiz. Spacing	Cylinder Diameter	Horiz. Spacing	Cylinder Diameter	Horiz. Spacing	Cylinder Diameter			
Over 5 Up to 10	4	3.5	8.0	2 in	8.0	2 in Note (2)	8.0	3 in	---	---	3x12
		7.0	9.0	2 in	9.0	2 in Note (2)	9.0	3 in			
		14.0	12.0	3 in	12.0	3 in	12.0	3 in			
Over 10 Up to 15	4	3.5	6.0	2 in	6.0	2 in Note (2)	6.0	3 in	---	3x12	---
		7.0	8.0	3 in	8.0	3 in	8.0	3 in			
		14.0	10.0	3 in	10.0	3 in	10.0	3 in			
Over 15 Up to 20	4	3.5	5.5	2 in	5.5	2 in Note (2)	5.5	3 in	3x12	---	---
		7.0	6.0	3 in	6.0	3 in	6.0	3 in			
		14.0	9.0	3 in	9.0	3 in	9.0	3 in			
Over 20	See Note (1).										

Footnotes to tables, and general notes on hydraulic shoring, are found in Appendix D, Item (g)
Note(1): See Appendix D, Item (g)(1)
Note(2): See Appendix D, Item (g)(2)
* Consult product manufacturer and/or qualified engineer for Section Modulus of available wales.

TABLE D - 1.4. ALUMINUM HYDRAULIC SHORING WALER SYSTEMS FOR SOIL TYPE C

Depth of Trench (Feet)	WALES		HYDRAULIC CYLINDERS						TIMBER UPRIGHTS		
	Vertical Spacing (Feet)	*Section Modulus (in(3))	Width of Trench (Feet)						Max. Horiz. Spacing (On Center)		
			Up to 8		Over 8 Up to 12		Over 12 Up to 15		Solid Sheet	2 ft	3 ft
			Horiz. Spacing	Cylinder Diameter	Horiz. Spacing	Cylinder Diameter	Horiz. Spacing	Cylinder Diameter			
Over 5 Up to 10	4	3.5	6.0	2 in	6.0	2 in Note (2)	6.0	3 in	3x12	---	---
		7.0	6.5	2 in	6.5	2 in Note (2)	6.5	3 in			
		14.0	10.0	3 in	10.0	3 in	10.0	3 in			
Over 10 Up to 15	4	3.5	4.0	2 in	5.0	2 in Note (2)	4.0	3 in	3x12	---	---
		7.0	5.5	3 in	5.5	3 in	5.5	3 in			
		14.0	8.0	3 in	8.0	3 in	8.0	3 in			
Over 15 Up to 20	4	3.5	3.5	2 in	3.5	2 in Note (2)	3.5	3 in	3x12	---	---
		7.0	5.0	3 in	5.0	3 in	5.0	3 in			
		14.0	6.0	3 in	6.0	3 in	6.0	3 in			
Over 20	See Note (1).										

Footnotes to tables, and general notes on hydraulic shoring, are found in Appendix D, Item (g)
Note(1): See Appendix D, Item (g)(1)
Note(2): See Appendix D, Item (g)(2)

- Consult product manufacturer and/or qualified engineer for Section Modulus of available wales.

4.H. – 29 CFR 1926 Subpart P Appendix E – Alternatives to Timber Shoring

- Part Number: 1926
- Part Title: Safety and Health Regulations for Construction
- Subpart : P
- Subpart Title: Excavations
- Standard Number: 1926.650
- Title: Alternatives to Timber Shoring.

ALTERNATIVES TO TIMBER SHORING

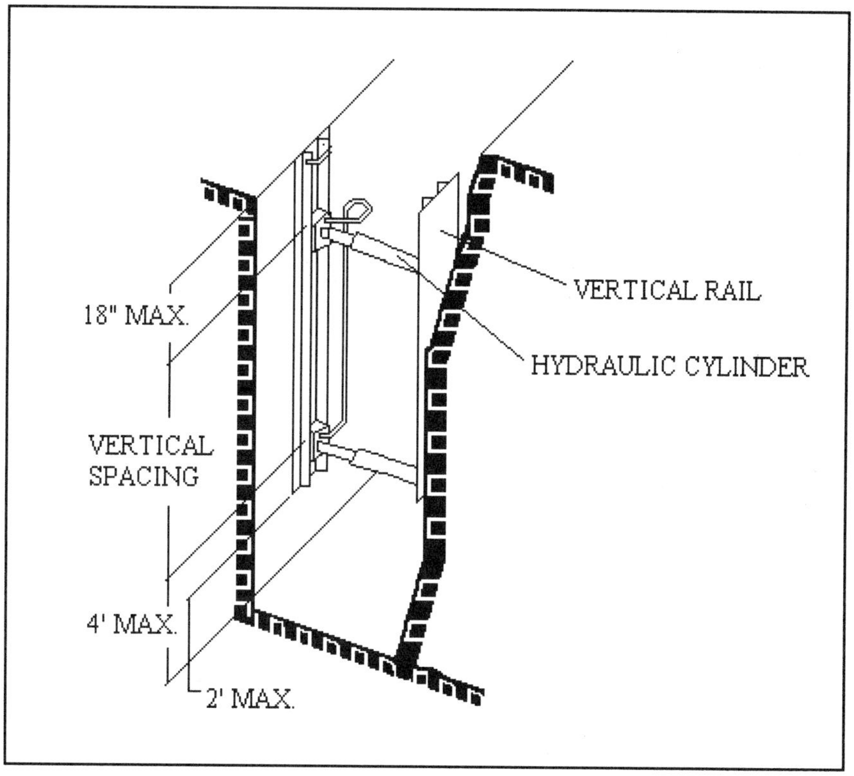

Figure No. 1 - Aluminum Hydraulic Shoring

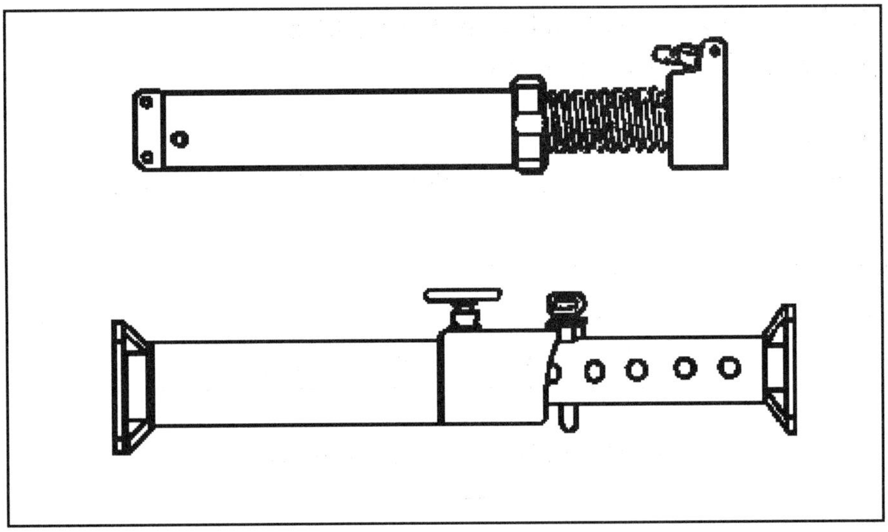

Figure No. 2 - Pneumatic/Hydralic Shoring

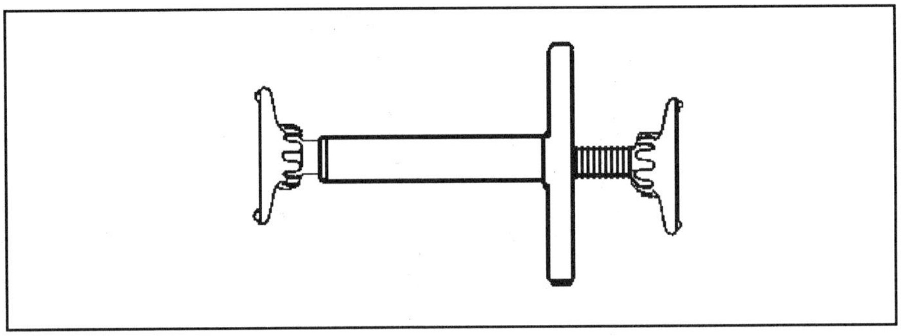

Figure No. 3 - Trench Jacks (Screw Jacks)

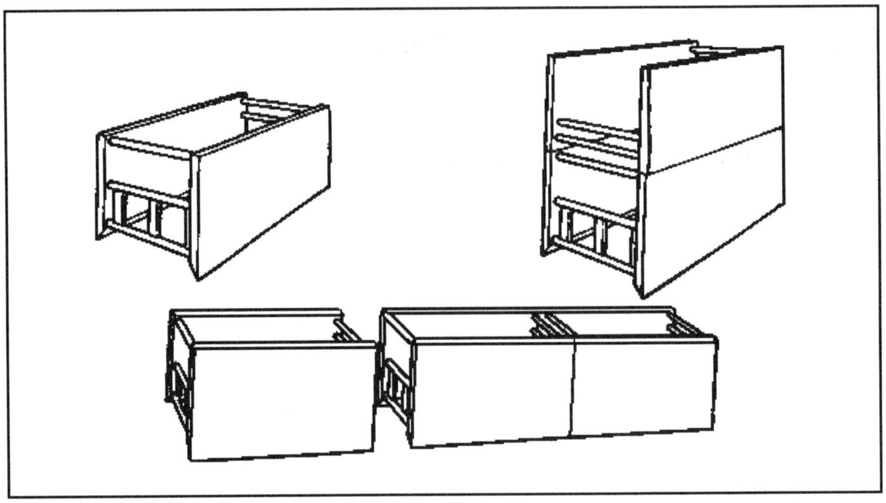

Figure No. 4 - Trench Shields

4.I. – 29 CFR 1926 Subpart P Appendix F – Selection of Protective Systems

- Part Number: 1926
- Part Title: Safety and Health Regulations for Construction
- Subpart : P
- Subpart Title: Excavations
- Standard Number: 1926.650
- Title: Selection of Protective Systems.

The following figures are a graphic summary of the requirements contained in subpart P for excavations 20 feet or less in depth. Protective systems for use in excavations more than 20 feet in depth must be designed by a registered professional engineer in accordance with 1926.652(b) and (c).

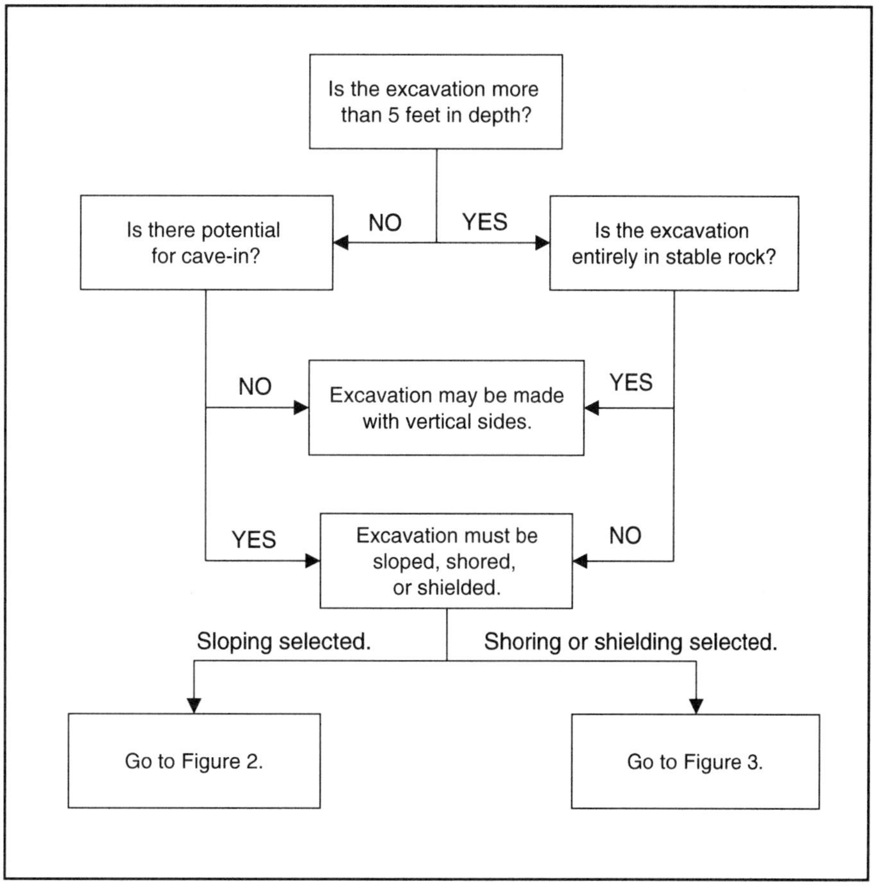

Figure 1 - Preliminary Decisions

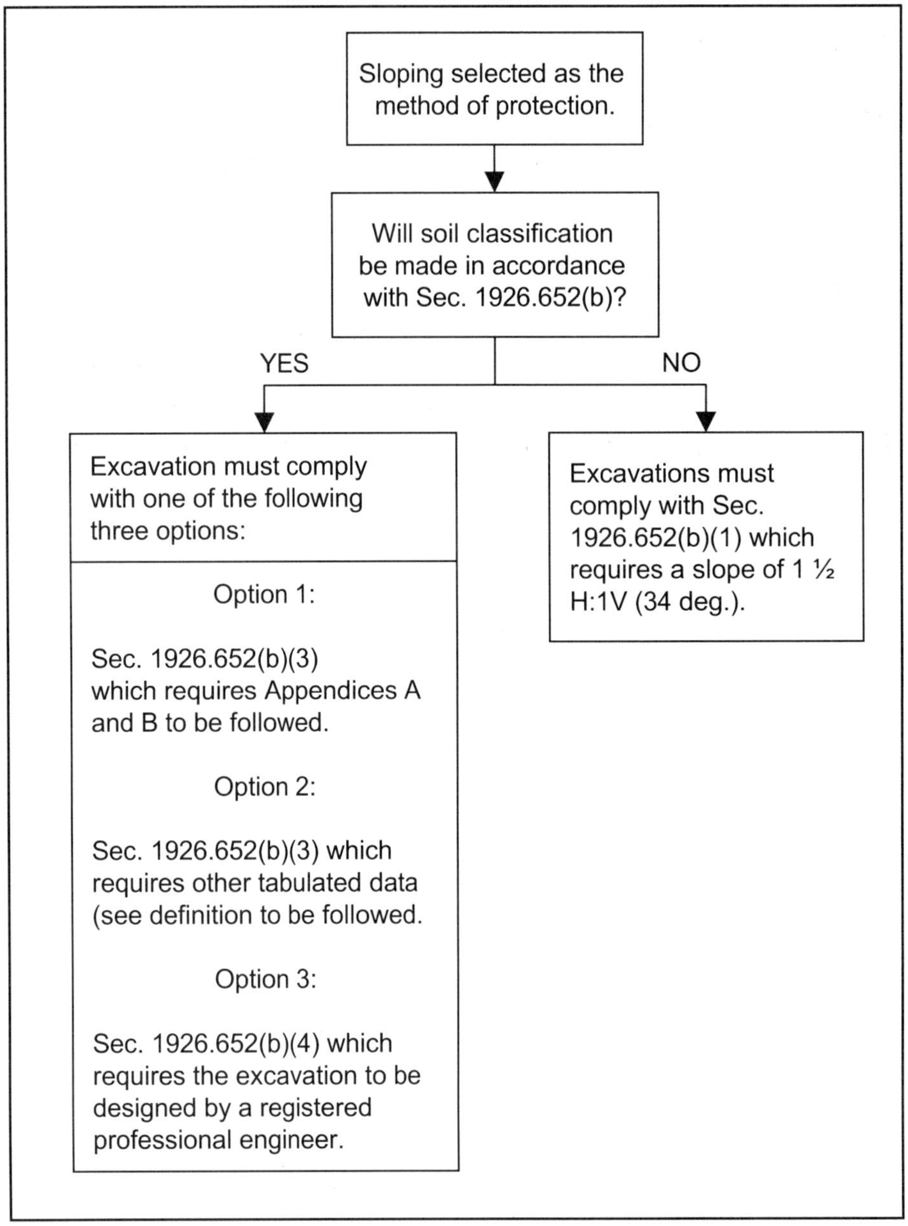

Figure 2 - Sloping Options

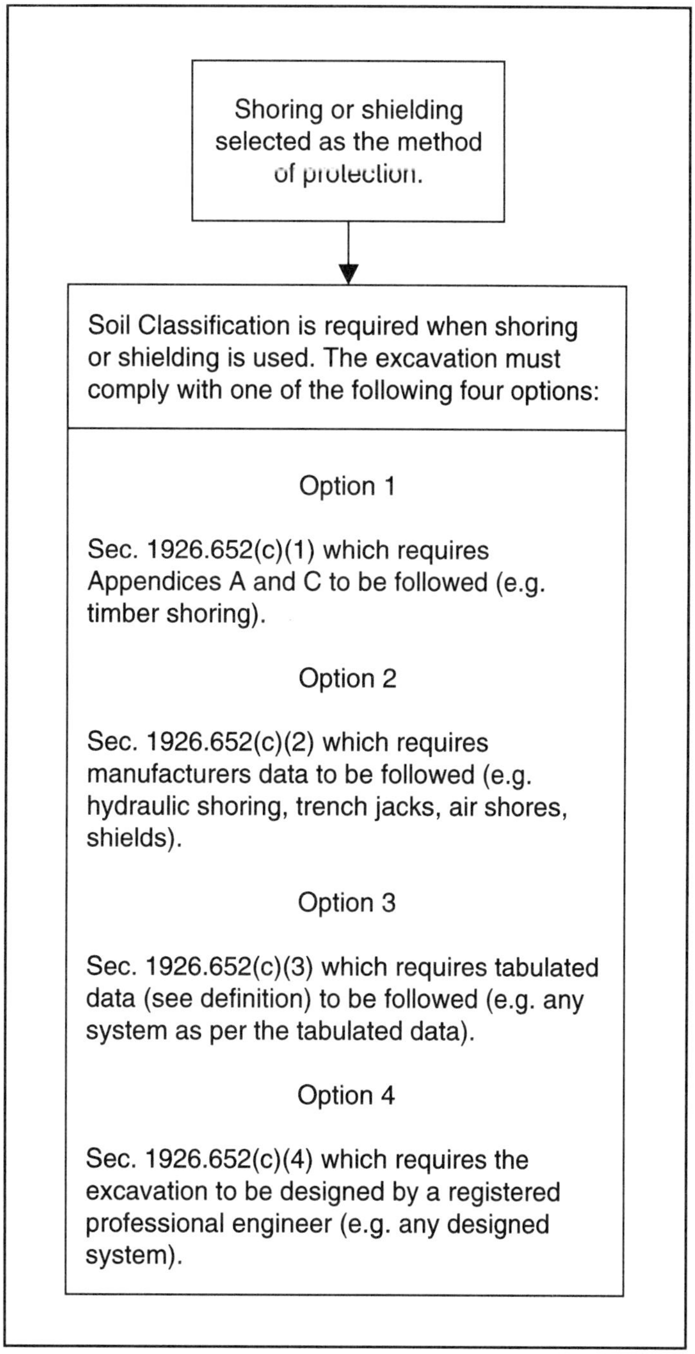

Figure 3 - Shoring and Shielding Options

Appendix A

Excavation Condition Log

All excavations 4 feet or deeper (or other excavations where employee exposure can be reasonably anticipated) shall be evaluated daily by a designated competent person before employees are permitted to enter the excavation for the first time. When the designated competent person deems another evaluation necessary during the same work shift, that evaluation shall be recorded on a separate excavation condition log.

Observation Date: _________ Time: _________ a.m./p.m. Location: _______________
Project: ____
Work Order Number: __________________ Job Supervisor: __________________
Phone #: _______________
Weather: ____

1. Measurements of excavation (in feet): Depth _____ Length _____ Width _____

2. Soil Type: Stable Rock ❐ Type A ❐ Type B ❐ Type C ❐ Same as Plan _________

3. Soil Classified: Visually ❐ Manually ❐

4. Soil Conditions: Dry ❐ Wet ❐ Stable ❐ Seeping/Saturated ❐ Sloughing ❐ Surface Tension Cracking ❐

5. Protective System(s): Benching ❐ Sloping ❐ Trench/Shield ❐ Timber Shoring ❐ Hydraulic Shoring ❐

6. Upper Bench: V_____H_______
Next Bench: V_____H_______
Additional Benches: V_____H_______ (V = Vertical, H = Horizontal)
V_____H_______

7. Is the lowest bench width twice its height? Yes ❐ No ❐ N/A ❐

8. Is excavated material stored 2 feet or more from edge of excavation? Yes ❐ No ❐ N/A ❐

9. Are there any signs of cracking in exposed walls of foundations? Yes ❐ No ❐ N/A ❐

10. Is there water seeping or running through the excavation? Yes ❐ No ❐ N/A ❐

11. Is shoring tight against sides? Yes ❐ No ❐ N/A ❐

12. Are shields placed according to requirements? Yes ❐ No ❐ N/A ❐

13. Is the protective system adequate? Yes ❐ No ❐ N/A ❐

14. Does the excavation require air monitoring? Yes ❐ No ❐ N/A ❐

15. Is a ladder or ramp within 25 feet of all workers? Yes ❐ No ❐ N/A ❐

16. Do walkways provided for crossing over excavations 6 feet or more above lower levels comply with OSHA 1926.502(b)? Yes ❐ No ❐ N/A ❐

17. What vehicles or equipment are operating in or near the excavation?

____

18. Is the excavation safe for personnel to enter? Yes ❐ No ❐

19. If "No," what corrective action(s) are required before entry?

____

20. Designated competent person signature: ______________________________

Appendix B

Site Assessment Questions

- ❒ During first and subsequent visits to an excavation site, the site's safety officer or other interested person may find the following questions useful.
- ❒ Is the cut, cavity, or depression a trench or an excavation?
- ❒ Is the cut, cavity, or depression more than 4 feet in depth?
- ❒ Is there water in the cut, cavity, or depression?
- ❒ Are there adequate means of access and egress?
- ❒ Are there any surface encumbrances?
- ❒ Is there employee exposure to vehicular traffic?
- ❒ Are adjacent structures stabilized?
- ❒ Does mobile equipment have a warning system?
- ❒ Is a competent person in charge of the operation?
- ❒ Is equipment operating in or around the cut, cavity, or depression?
- ❒ Are procedures required to monitor, test, and control hazardous atmospheres?
- ❒ Did a competent person determine soil type?
- ❒ Was a soil testing device used to determine soil type?
- ❒ Are the spoils placed 2 feet or more from the edge of the cut, cavity, or depression?
- ❒ Is the depth 20 feet or more for the cut, cavity, or depression?

- ❒ Has a registered professional engineer approved the procedure if the depth is more than 20 feet?
- ❒ Does the procedure require benching or multiple benching? Shoring? Shielding?
- ❒ If provided, do shields extend at least 18 inches above the surrounding area if it is sloped toward the excavation?
- ❒ If shields are used, is the depth of the cut more than 2 feet below the bottom of the shield?
- ❒ Are any required surface crossings of the cut, cavity, or depression the proper width and fitted with handrails?
- ❒ Are means of egress from the cut, cavity, or depression no more than 25 feet from the work?
- ❒ Is emergency rescue equipment required?
- ❒ Is there documentation of the minimum daily excavation inspection?

Appendix C

Additional Resources

OSHA

www.osha.gov

OSHA plays a vital role in preventing on-the-job injuries and illnesses through outreach, education, and compliance assistance. Most recently the agency has provided a broad array of training and informational materials on its recordkeeping standard, as well as materials to assist employers and workers in understanding and complying with its new steel erection standard.

OSHA offers an extensive website (www.osha.gov) with nearly 450,000 pages of information. It includes a special section devoted to assisting small businesses as well as interactive e-tools to help employers and employees. Free workplace consultations are available in every state to small businesses that want onsite help in establishing safety and health programs and identifying and correcting workplace hazards. In addition, OSHA has a network of Compliance Assistance Specialists in local offices available to provide tailored information and training to employers and employees.

National Audio Visual Center

www.ntis.gov/products/types/audiovisual/index.asp

Training is available in various programs such as the one available from the National Audio Visual Center (NAC) (Order No. 689601) titled *Safety and Health Excavation and Trenching Operations.* This is an instructional program designed to increase awareness and understanding of the problems and hazards in excavation and trenching operations. It includes an instructor's guide and 139 slides.

Another NAC program is *Trenching* (Order No. 007516), a slide-tape hazard recognition program that includes 96 slides, instructor's guide, workbook, and course outline. Also available from NAC is a one-day instructional program, *Sloping, Shoring and Shielding* (Order No. 009863), that involves classroom sessions and a hands-on workshop. This package includes an instructor's manual, outline for field exercise/ workshop, and 60 slides. More information on training can be found in Chapter 4.

Underground Focus: The Magazine of Below-Ground Damage Prevention

www.underspace.com

Commercial and industrial knowledge and experience should be obtained from available sources. One excellent source is the Internet site for Underground Focus: The Magazine of Below-Ground Damage Prevention. This site provides extensive information on recent accidents, accident bulletins, and accident photos. It also includes a call-before-you-dig directory, descriptions of detection instruments, a network of damage prevention professionals, and links to other relevant sites.

Center for Subsurface Strategic Action

Another way to use commercial knowledge and experience is to become associated with industry trade groups such as the Center for Subsurface Strategic Action (CSSA) and the National Utility Locating Contractors Association (NULCA). More information on these organizations can be obtained from the Underground Focus website, or by calling 800-435-4869 or 715-635-6004.

American Society of Civil Engineers

www.asce.org

The American Society of Civil Engineers (ASCE) has formed a national standards group titled "Collection and Depiction of Existing Utility Data on Design and Construction Documents."

Appendix D

Questions on Excavation Safety

1. ____________________ means a method of protecting employees from cave-ins by excavating the sides of an excavation to form one or a series of horizontal levels or steps, usually with vertical or near-vertical surfaces between levels.
   a. Shoring
   b. "Benching (Benching system)"
   c. Protective system
   d. "Shield (Shield system)"

2. ____________ means the separation of a mass of soil or rock material from the side of an excavation, or the loss of soil from under a trench shield or support system, and its sudden movement into the excavation, either by falling or sliding, in sufficient quantity so that it could entrap, bury, or otherwise injure and immobilize a person.
   a. Failure
   b. Cave-in
   c. Kickout
   d. Unstable

3. _________________ means the members of a shoring system that retain the earth in position and in turn are supported by other members of the shoring system.
   a. Shielding
   b. Cross braces
   c. Sheeting
   d. Support system

4. ________________ means a method of protecting employees from cave-ins by excavating to form sides of an excavation that are inclined away from the excavation so as to prevent cave-ins.
   a. Sloping (sloping system)
   b. Benching
   c. Shoring
   d. Shielding

5. All ____________ that are located so as to create a hazard to employees shall be removed or supported, as necessary, to safeguard employees.
   a. Ramps
   b. Support systems
   c. Cross braces
   d. Surface encumbrances

6. ________________ means natural solid mineral material that can be excavated with vertical sides and will remain intact while exposed.
   a. Stable rock
   b. Wales
   c. Unstable rock
   d. Spoils

7. Structural ramps that are used solely by employees as a means of access or egress from excavations shall be designed by a ________________.
   a. Structural engineer
   b. Qualified person
   c. Trained employee
   d. Competent person

8. Adequate precautions shall be taken to prevent employee exposure to atmospheres containing less than _______ percent oxygen and other hazardous atmospheres.
   a. 17.9
   b. 18.9
   c. 19.5
   d. 20.9

9. While the excavation is open, ____________ shall be protected, supported, or removed as necessary to safeguard employees.
   a. Underground installations
   b. Side walls
   c. Equipment
   d. Spoils

10. ______ soil is cohesive soil with an unconfined compressive strength greater than .5tsf, and includes soil such as silt, silt loam, and sandy loam.
   a. Type A
   b. Type B
   c. Type C
   d. Stable rock

11. A competent person is required to inspect excavations for hazardous conditions _________.
   a. At the end of the work shift
   b. Before employees enter the excavation and as conditions change
   c. At least once a day
   d. When the job is completed

12. All excavations greater than _______ feet in depth with actual or potential atmospheric hazards require testing.
   a. 2
   b. 4
   c. 10
   d. 20

13. Spoils must be placed at least 2 feet from the _________ of the excavation:
   a. Bottom
   b. Middle
   c. Nearest edge
   d. Top

14. Safe means of access is required for employees working in trench excavations __________ feet deep or more.
    a. 4
    b. 5
    c. 8
    d. 20

15. The greatest risk in trenching and excavation work, and one of primary concern is ___________.
    a. Atmospheric hazards
    b. Cave-ins
    c. Fall hazards
    d. Electrocution

Answers

1. ____________________ means a method of protecting employees from cave-ins by excavating the sides of an excavation to form one or a series of horizontal levels or steps, usually with vertical or near-vertical surfaces between levels.

   a. Shoring

   b. "Benching (Benching system)"

   c. Protective system

   d. "Shield (Shield system)"

2. ____________ means the separation of a mass of soil or rock material from the side of an excavation, or the loss of soil from under a trench shield or support system, and its sudden movement into the excavation, either by falling or sliding, in sufficient quantity so that it could entrap, bury, or otherwise injure and immobilize a person.

   a. Failure

   b. Cave-in

   c. Kickout

   d. Unstable

3. ________________ means the members of a shoring system that retain the earth in position and in turn are supported by other members of the shoring system.

   a. Shielding

   b. Cross braces

   c. **Sheeting**

   d. Support system

4. __________________ means a method of protecting employees from cave-ins by excavating to form sides of an excavation that are inclined away from the excavation so as to prevent cave-ins.

   a. **Sloping (sloping system)**

   b. Benching

   c. Shoring

   d. Shielding

5. All ___________ that are located so as to create a hazard to employees shall be removed or supported, as necessary, to safeguard employees.
   a. Ramps
   b. Support systems
   c. Cross braces
   d. Surface encumbrances

6. ________________ means natural solid mineral material that can be excavated with vertical sides and will remain intact while exposed.
   a. Stable rock
   b. Wales
   c. Unstable rock
   d. Spoils

7. Structural ramps that are used solely by employees as a means of access or egress from excavations shall be designed by a __________________.
   a. Structural engineer
   b. Qualified person
   c. Trained employee
   d. Competent person

8. Adequate precautions shall be taken to prevent employee exposure to atmospheres containing less than _______ percent oxygen and other hazardous atmospheres.
   a. 17.9
   b. 18.9
   c. 19.5
   d. 20.9

9. While the excavation is open, _____________ shall be protected, supported, or removed as necessary to safeguard employees.
   a. Underground installations
   b. Side walls
   c. Equipment
   d. Spoils

10. ______ soil is cohesive soil with an unconfined compressive strength greater than .5tsf, and includes soil such as silt, silt loam, and sandy loam.
    a. Type A
    b. **Type B**
    c. Type C
    d. Stable rock

11. A competent person is required to inspect excavations for hazardous conditions ________.
    a. At the end of the work shift
    b. **Before employees enter the excavation and as conditions change**
    c. At least once a day
    d. When the job is completed

12. All excavations greater than ______ feet in depth with actual or potential atmospheric hazards require testing.
    a. 2
    b. 4
    c. 10
    d. 20

13. Spoils must be placed at least 2 feet from the ________ of the excavation:
    a. Bottom
    b. Middle
    c. **Nearest edge**
    d. Top

14. Safe means of access is required for employees working in trench excavations ________ feet deep or more.
    a. 4
    b. 5
    c. 8
    d. 20

15. The greatest risk in trenching and excavation work, and one of primary concern is ___________.
    a. Atmospheric hazards
    b. Cave-ins
    c. Fall hazards
    d. Electrocution

Appendix E

One-Call Systems International Directory

2002 Excavator's Damage Prevention Guide

reprinted with permission from the

American Public Works Association

Contents

EXCAVATOR DAMAGE PREVENTION GUIDE AND ONE-CALL SYSTEMS DIRECTORY

INTRODUCTION

This guide is intended as a resource to aid in the prevention of damage to underground facilities. It includes the American Public Works Association Color Code and marking guidelines, a listing of the known international One-Call Centers, and a summary of damage prevention laws. The laws indicate that it is the responsibility of the excavator to notify the one-call center before excavation occurs.

Coordination, cooperation and communication are what one-call systems are all about. A one-call system is a tool used in preventing facility dig-ups. It is a communication link between excavators and buried-plant owners and operators.

One-call is a safety program designed to cut the cost of pipe and cable repairs and diminish the potential hazard posed to workers and the general public whenever excavation is done.

Please use this guide as a training and information tool. The one-call center nearest you can also serve as a valuable resource for your training and safety programs. Please contact them for assistance. Their number is listed in the One-Call Directory. Most one-call centers have people available for presentations and have an excellent selection of promotional materials.

CALL BEFORE YOU DIG!

NORTH AMERICAN ONE-CALL REFERRAL SYSTEM

Dig Safely.

1-888-258-0808

This toll-free number can be accessed 24 hours per day, every day of the year, from anywhere in the United States or Canada. It is a mechanical service, allowing callers to obtain one-call center telephone numbers for the state or provinces where excavation activity sites are located.

OCSI MISSION STATEMENT

The purpose of One-Call Systems International (OCSI) is to promote facility damage prevention and infrastructure protection through education, guidance and assistance.

DIRECTORY DISCLAIMER

This directory is not all-inclusive, nor is each system listed necessarily a member of One-Call Systems International or an adherent to its principles. The purpose of the directory is to illustrate the extent of one-call service available. The accuracy of information is not guaranteed by APWA or the one-call systems, and there well may be inaccuracies or changes at or after publication. It should be clearly understood that each system provides a different level and type of service. Any person using this directory is responsible for verifying information, including the extent and limit of such service, inter-alia, from local source or as required by state law or other regulation, or as required by the local system.

ONE-CALL SYSTEM DEFINITION

What Is One-Call?

One-Call is a communication system established by two or more utilities, government agencies, or other operators of underground facilities to provide one telephone number for excavating, tunneling, demolition, or any other similar work. This one-call system provides the participating members an opportunity to identify and locate their underground facilities.

Why Is It Needed?

Damage to underground facilities increased considerably following the building boom of the 1950s, 60s, and early 70s when the trend was to go underground with utilities. Thousands of miles of underground facilities were vulnerable to excavating machines, such as backhoes, and the resulting damage interrupted utility service and threatened life, health and property. The use of one-call systems and location of underground facilities prior to excavation reduce the risk of damage or service interruptions.

Excavator Responsibilities

1. Notify one-call center in advance of work in accordance with the local regulations. (See state summary in the following One-Call Directory)
2. Meet with owner's field representative if necessary.
3. Look for, observe and preserve marking.
4. Excavate in a careful and prudent manner and in accordance with local regulations. (See state summary in the following One-Call Directory)
5. Notify facility owner immediately of any damage to facility, coating or special backfill—DO NOT COVER UP.

Facility Owners Responsibilities

1. Actively participate in the local one-call center.
2. Respond to notices in a timely manner, in accordance with one-call center guidelines and local regulations.
3. Properly mark facility location, in accordance with local regulations. (See Color Code below)
4. Honor all field meeting appointments with excavators.
5. Respond promptly to call for assistance or information.

APWA Uniform Color Code

(uses ANSI standard Z535.1 Safety Color Code)

WHITE	Proposed Excavation
PINK	Temporary Survey Markings
RED	Electric Power Lines, Cables, Conduit and Lighting Cables
YELLOW	Gas, Oil, Steam, Petroleum or Gaseous Materials
ORANGE	Communication, Alarm or Signal Lines, Cables or Conduit
BLUE	Potable Water
PURPLE	Reclaimed Water, Irrigation and Slurry Lines
GREEN	Sewers and Drain Lines

GUIDELINES FOR UNIFORM TEMPORARY MARKING OF UNDERGROUND FACILITIES

This marking guide provides for universal use and understanding of the temporary marking of subsurface facilities to prevent accidents and damage or service interruption by contractors, excavators, utility companies, municipalities or any others working on or near underground facilities.

ONE-CALL SYSTEMS

The One-Call damage prevention system shall be contacted prior to excavation.

PROPOSED EXCAVATION

Use white marks to show the location, route or boundary of proposed excavation. Surface marks on roadways do not exceed 1.5" by 18" (40 mm by 450 mm). The facility color and facility owner identity may be added to white flags or stakes.

USE OF TEMPORARY MARKINGS

Use color-coded surface marks (i.e., paint or chalk) to indicate the location or route of active and out-of-service buried lines. To increase visibility, color-coded vertical markers (i.e., stakes or flags) should supplement surface marks. Marks and markers indicate the name, initials or logo of the company that owns or operates the line, and width of the facility if it is greater than 2" (50 mm). Marks placed by other than line owner/operator or its agent indicate the identity of the designating firm. Multiple lines in joint trench are marked in tandem. If the surface over the buried line is to be removed, supplementary offset markings are used. Offset markings are on a uniform alignment and clearly indicate the actual facility is a specific distance away.

TOLERANCE ZONE

Any excavation within the tolerance zone is performed with non-powered hand tools or non-invasive method until the marked facility is exposed. The width of the tolerance zone may be specified in law or code. If not, a tolerance zone including the width of the facility plus 18" (450 mm) measured horizontally from each side of the facility is recommended.

ADOPT UNIFORM COLOR CODE

The American Public Works Association encourages public agencies, utilities, contractors, other associations, manufacturers and all others involved in excavation to adopt the APWA Uniform Color Code, using ANSI standard Z535.1 Safety Colors for temporary marking and facility identification.

Rev. 4/99

INSTRUCTIONS FOR RECOMMENDED MARKINGS OF UNDERGROUND UTILITIES-Approved by APWA Board of Directors - 9/12/2001

1. Utility lines will be indicated by markings using current APWA color codes. Markings should be 18"-24" in length and 2" in width.

2. The owner of a facility should be indicated by initials or by name in letters 6" high at the beginning and end of the locate. On long locates the facility owner should be indicated every 100'.

3. When known, the total number of lines within the ground will be indicated. The number of lines indicated should be based on the physical lines "that you could place your hands on." Multiple cables twisted together to form a single facility, as in the case of electric lines, would be considered one cable for locate purposes.

4. If a facility is known to be present but the total number of lines for a facility cannot be determined a corridor marker may be used. The corridor marker should indicate the approximate width of the facility. A marking resembling the letter "H" lying on its side will indicate the corridor marker.

5. When known, the size of the line being located will be indicated. Line size will indicate the outside diameter of the pipe or structure. The oversized utility marking should indicate the approximate size of pipe or structure. A mark resembling the letter "H" lying on its side, bisected by line extending along its length will indicate the oversized utility marking. The committee discussions centered on indicating sizes of single physical structures such as gas lines, sewer lines, water lines, and storm drains.

6. Duct structures, whether a single duct or multiple ducts, will be indicated by duct symbol indicating the approximate width of the duct structure. The duct marker will be indicated by a marking resembling a diamond bracketed by two parallel lines. The committee did not state whether indicating size should extend to indicating size of duct structures (telecommunications, electric) or indicating pair count for telephone lines.

7. When known, the pressure of a gas facility will be indicated. Gas pressure will be indicated as either low pressure, intermediate high pressure or high pressure. The committee did not determine whether gas pressure would be indicated in instances of high pressure only or for all pressures. The committee did not determine whether the type of pipe should be indicated (plastic, steel, cast iron, etc.).

8. When known, termination points, dead ends and stub outs should be indicated. The committee reviewed the National Utility Locating Contractors Association (NULCA) recommendation, which resembles the letter "T" with drop downs.

9. When there is a strong likelihood that marks may be destroyed, offsets should be used. Offsets are indicated on a permanent surface and are placed parallel to the running line of the facility. The offset should indicate the distance from the offset to the facility and should identify the facility owner and if necessary size of the facility.

Line Markings

Corridor Markings

Oversized Utility Markings

Conduit Markings

Marking Gas Lines-High Pressure

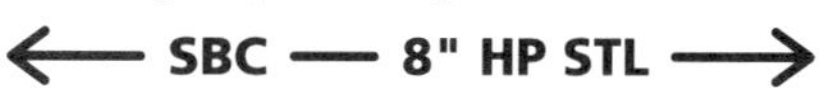

Marking Termination Point, Dead End, Stub Outs

Marking Offsets

No Conflict
(No utilities within the requested area)

NO/MCI NO/EPG

Proposed Markings - Consesus not reached by committee

Electronic Marker

EM

Marking Buried Splices, Valves, Manholes

One-Call Systems International (OCSI) 2002 Members

ONE-CALL SYSTEMS INTERNATIONAL (OCSI) 2002-2003 OFFICERS

Chair:

James D. Maniscalco
Executive Director
Utility Notification Center of CO
12600 W Colfax Ave, Ste B310
Lakewood, CO 80215
Phone: 303-205-6301
Fax: 303-234-1712
E-mail: jdman@uncc.org

Vice Chair:

Sandra Holmes
Executive Director
Arizona Blue Stake, Inc
4415 S Wendler Dr, Ste 105
Tempe, AZ 85282
Phone: 602-659-7500
Fax: 602-659-7510
E-mail: smholmes@azbsinc.com

Secretary:

Gary Craig
Vice President
One Call Systems, Inc
8227 Middlebury Ln
Houston, TX 77070-3618
Phone: 281-970-0505
Fax: 281-970-0110
E-mail: gcraig@1-call.com

Legal Advisor:

William Boswell
Partner
McGuireWoods
CNG Tower
625 Liberty Ave, 23rd Fl
Pittsburgh, PA 15222
Phone: 412-667-7901
Fax: 412-667-7952
E-mail: wboswell@mcguirewoods.com

Jean Abi-aoun
Seminole County
(407) 426-8994
Fax: (407) 426-8977
jabiaoun@co.seminole.fl.us

Sharon Able
Kentucky Underground Protection, Inc
(877) 711-6225
Fax: (502) 633-0710
kupi@bellsouth.net

Richard Abraham
Kansas One Call System, Inc
(316) 262-4270

Lisa M. Adams
City of Lincoln
(402) 441-6855
Fax: (402) 441-8493
ladams@ci.lincoln.nc.us

Wayne L. Babcock
Diggers Hotline of Nebraska
(402) 731-5858
Fax: (402) 334-7640

Charlotte A. Baclesse
Missouri One Call System
(573) 556-8100
Fax: (573) 635-8402

Rodney Bacon
Password
(509) 624-5235
Fax: (509) 624-0220

Arnie E. Bailey
Enron Transportation & Storage
(505) 625-8005
Fax: (505) 260-0968

William E. Balin
Pennsylvania One Call System, Inc
(412) 464-7102
Fax: (412) 464-7165
webalin@pa1call.org

Kurt Baltrusch, P.E.
Utilities Underground Location Ctr
(425) 454-6888
Fax: (425) 451-2385

Bobbie Bennett
Arizona Blue Stake, Inc
(602) 659-7500
Fax: (602) 659-7510

Tonya Bethke
One Call Concepts, Inc
(651) 454-8388
Fax: (651) 454-0170
TonyaBethke@isd.net

Lawrence Bishop
Dig Safely New York, Inc
(315) 437-7394
Fax: (315) 437-2621

Paul Blouin
Louisiana One Call
(225) 275-3700
Fax: (225) 272-1967
paul@laonecall.com

Tim Boatfield
Utilities Protection Center
(770) 623-4332
Fax: (770) 623-1847
tboatfield@gaupc.com

Ben Booth, PRC
Iowa One Call
(515) 278-8700
Fax: (515) 278-0245
onecall@iowaonecall.com

William Boswell
McGuireWoods
(412) 667-7901
Fax: (412) 667-7952
wboswell@mcguirewoods.com

Darab Bouzarjomehri
Project Design Consultants
(619) 881-3433
Fax: (619) 234-0349
darab@project design.com

Fiona Bowen
Utilities Protection Center
(770) 623-4332x204
Fax: (770) 623-1847

Chris Brown
New Mexico One Call
(505) 260-1165
Fax: (505) 260-0968

Wendy Bullins
North Carolina One Call
(336) 855-5760
Fax: (336) 299-1914

Daniel W. Buol
Damage Prevention Consulting
(863) 859-0022
Fax: (863) 815-1113
dbuol@aol.com

David Burstein
PSMJ Resources Inc
(770) 723-9651
Fax: (815) 461-7478
dburstein@psmj.com

Barbara Bye
One Call Systems, Inc
(732) 394-3010
Fax: (732) 394-3011
bbye@i-2000.com

Ralph Byrd
Jackson Emc
(770) 822-3234

Claudette L. Campbell
Utilities Protection Center, Inc.
(770) 623-4332, ext 200
Fax: (770) 623-1847
ccampbell@gaupc.com

Carolyn Carter
North Carolina One Call
(336) 855-5760
Fax: (336) 299-1914
carolyn@ncocc.org

Robert Chisholm, P Eng
Alberta One Call Corp
(403) 531-3712
Fax: (403) 531-3703
bchisholm@alberta1call.com

Deborah Clark
C&B Associates, Inc
(940) 682-4213
Fax: (940) 682-7309
dclark@candb2.com

Salvatore Colella
MISS DIG System, Inc
(248) 874-3401
Fax: (248) 874-3410

Roger Cox
Arkansas One-Call Center
(501) 328-2500
Fax: (800) 482-7997

Gary W. Craig
One Call Systems, Inc
(281) 970-0505
Fax: (281) 970-0110
gcraig@1-call.com

Adrien F. Criner
Utilities Underground Loc
(406) 442-3070
Fax: (406) 443-3554
uulcrep@msn.com

Judy Dahl
Utility Notification Center of CO
(303) 985-1581
Fax: (303) 985-0680
grawsd@msn.com

Barry Davis
City of Burnaby
(604) 294-7464
Fax: (604) 294-7425
davis@city.burnaby.bc.ca

Frank D. DeCelles
City of Plantation
(954) 452-2536
Fax: (954) 452-2548
pw@gate.net

Brenda E. Derrig
City of Fargo
(701) 241-1549
Fax: (701) 241-8101

William E. Dexheimer
Missouri One Call System
(573) 556-8100
Fax: (573) 635-8402

Richard E. Dunkerly
Utilities Protection Center
(770) 623-4332
Fax: (770) 623-1847
dunkerly@ix.netcom.com

Larry Englerth
South Dakota One Call
(605) 399-0529
Fax: (605) 339-0529
sdonecall@qwest.net

Dale Enoch
Arkansas One-Call System, Inc.
(501) 328-2500
Fax: (501) 328-2522
denoch@arkonecall.com

Jose Espino
Utility Notification Center of CO
(303) 205-6301
Fax: (303) 234-1712

Gary Evans
Waukesha County WI
(262) 548-7740
Fax: (262) 896-8097
bkalister@co.waukesha.wi.us

Donald R. Evans
Operating Committee-UULC
(877) 668-4001, ext 3307
Fax: (360) 871-8192
evans_dr@msn.com

Mohammed Fakhrriddine
San Diego County CA, Solid Waste
(858) 694-2822
Fax: (858) 694-2499
mrakhrpw@com.san.diago.com

Jeff Farrells
Utility Notification Center of CO
(719) 520-4401
Fax: (719) 520-4318
jeffrey.farrells@coastalcorp.com

Holly Files
Utilities Protection Center
(770) 623-4332
Fax: (770) 623-1847

Gregory Fisch
Julie, Inc
(815) 741-5003
Fax: (815) 741-5048
fisch@julie1call.com

Norman Fitzgerald
Fairfax County Water Authority
(703) 289-6334
Fax: (703) 289-6382

Anita Floyd
Utilities Underground Location Ctr
(503) 242-5138
Fax: (503) 242-8729

Todd Foje
Diggers Hotline of Nebraska
(402) 778-3024
Fax: (402) 334-7640

Rick E. Forner
Wight Consulting Engineers, Inc
(847) 381-1800
Fax: (847) 381-1875
gwight@aol.com

Bob Foster
Dig Safely New York, Inc
(315) 437-7394
Fax: (315) 437-2621
rfoster@digsafelynewyork.com

Robert E. Foster
Dig Safely New York, Inc
(315) 437-7394
Fax: (315) 437-2621
rfoster@ufpo.org

Kathleen A. Fournier
MISS DIG System, Inc
(248) 874-3401
Fax: (248) 874-3410
kafournier@mindspring.com
kafournier@missdig.org

Terry R. Fox
Iowa One Call
(515) 281-2316
Fax: (515) 242-4398

David T. Frey
Louisiana One Call System, Inc.
(225) 275-3700
Fax: (225) 272-1967

Mark A. Frost
JULIE, Inc
(815) 741-5000
Fax: (815) 741-5958
frost@julie1call.com

James G. Gerblick
Underground Technology, Inc
(805) 658-7801

John L. German, P.E.
PBS&J
(210) 828-9494
Fax: (210) 828-7282
jlgerman@pbsj.com

William R. Glass
(205) 680-5625
Fax: (205) 680-4400
ugocglass@aol.com

George L. Glenn
North Carolina One Call Center, Inc.
(336) 855-5760
Fax: (336) 299-1914
george@ncocc.org

Donald H. Gordon
(414) 421-3749
Fax: (414) 421-0973
__D@email.msn.com

Fran Gough
Utilities Underground Location Ctr
(425) 235-3034
Fax: (425) 235-3098

Bruce Gould
Utilities Underground Location Ctr
(253) 862-8602
Fax: (253) 862-8538

Ciro Guastella
One Call Users' Council, Inc
(718) 275-3135
Fax: (718) 631-6723
nycocuc@aol.com

Gary Hajek
Cross Valley Water District
(360) 668-6766
Fax: (360) 668-9634

W. Gary Hansen
Blue Stakes of Utah
(801) 539-8683
Fax: (801) 530-0562
garyh@bluestakes.org

Bill Heath
Sunshine State One-Call of Florida
(386) 575-2001
Fax: (386) 575-2032

Scott P. Henley
Henley Consulting, Inc
(604) 257-1900
Fax: (604) 451-0344
henley@islandnet.com

Richard G. Hewitt
Transportation Utilities Public Work
(613) 580-2424 x25658
Fax: (613) 580-4768
Richard.hewitt@city.ottawa.on.ca

Allan Heydorn
Cygnus Business Media
(920) 563-6388
Fax: (920) 563-1699
allan.heydorn@cygnuspub.com

Michael W. Heyer
Underground Service Alert
(925) 798-9504
Fax: (925) 798-9506
mheyer@usan.org

Donald B. Heyer
USA North
(925) 798-9504 x 4
Fax: (925) 798-1683
dheyer@usan.org

Thomas Hoff, Sr.
One Call Concepts, Inc
(410) 712-0082
Fax: (410) 712-0838

John D. Hohol
Berntsen International
(608) 249-8549
Fax: (800) 249-9794
jhohol@berntsen.com

Lisa Holder
Pennsylvania One Call System, Inc
(412) 464-7111
Fax: (412) 464-7104

Sandra M. Holmes
Arizona Blue Stake, Inc
(602) 659-7500 x 2223
Fax: (602) 659-7510
smholmes@azbsinc.com

James L. Holzer
One Call Concepts, Inc
(301) 621-1317
Fax: (301) 621-1711
jamesl@erols.com

Merle Hoppenworth
Iowa One Call
(515) 278-8700
Fax: (515) 278-0245

Mike Horn
One Call Concepts, Inc
(813) 549-4848
Fax: (813) 549-6233

Mark T. Horn
Town of Gilbert
(480) 503-6420
Fax: (480) 503-6404
markh@ci.gilbert.az.us

Mark Jadoul
Arizona Blue Stake, Inc
(602) 659-7500 x2230
Fax: (602) 659-7510
msjadoul@azbsinc.com

Joe Jainga
Utilities Underground Location Ctr
(425) 454-6888
Fax: (425) 451-2385
jjaing@exch.puget.com

Joseph Jainga
Utilities Underground Location Ctr
(206) 224-2120
Fax: (206) 224-2318
jjaing@exch.puget.com

Kim L. Jenkins
Alabama Line Location Center
(205) 414-1840
Fax: (205) 414-1899
kjenk@ix.netcom.com

Dawn Jester
One Call Concepts, Inc
(316) 687-2102
Fax: (316) 687-3753

Douglas R. Johns
Montana One Call Center
(406) 752-6811
Fax: (406) 752-7306
djohns@centurytel.net

Brian D. Johnson
Vibco, Inc
(800) 633-0032
Fax: (401) 539-2584

Sammy P. Johnson
Mississippi One Call System, Inc
(601) 362-4322
Fax: (601) 366-7666
ms1call@bellsouth.net

Mio Johnson
Municipal Light & Power
(907) 263-5408
Fax: (907) 263-5804
johnsonms@ci.anchorage.ak.us

David Jones
Missouri One Call System
(573) 556-8100
Fax: (573) 635-8402

David U. Jones
Missouri One Call System
(573) 635-1818
Fax: (573) 635-8402
dave@occinc.com

Jon Jones
New Mexico One Call
(505) 241-4571
Fax: (505) 241-2392

Douglas H. Kenyon
Diggers Hotline of Nebraska
(402) 731-5858
Fax: (402) 334-7640

Khrysanne Kerr
Ohio Utilities Protection Service
(800) 362-2764
Fax: (330) 545-7245
khrysanne@att.net

William G. Kiger
Pennsylvania One Call System, Inc
(412) 464-7111
Fax: (412) 464-7104
bgkiger@pa1call.org

Ellen M. Kiger
Pennsylvania One Call System, Inc
(412) 464-7111
Fax: (412) 464-7104
emkiger@pa1call.org

Kirk Kirkpatrick
Pennsylvania One Call System, Inc
(412) 464-7111
Fax: (412) 464-7104

James Kovach
Miss Utility of West Virginia, Inc.
(304) 345-3959
Fax: (304) 345-1851

Chad Krueger
Diggers Hotline
(414) 259-0676
Fax: (414) 259-1453

Patricia A. Lama
Portland General Electric
(503) 570-4400
Fax: (503) 570-4420
patti_lama@pgn.com

John Lansford
Missouri One-Call System, Inc.
(573) 6356-1818
Fax: (573) 635-8402
johnl@occinc.com

Thelma Latham
One Call Concepts, Inc
(225) 275-3700
Fax: (225) 272-2770
thelma@laonecall.com

Norman J. Launi
Miss Utility of West Virginia
(304) 446-5513
launinj@attmail.com

Laraine Lawson
Alberta One Call Corp
(403) 531-3712
Fax: (403) 531-3703

Lawrence O. Lee
Cincinnati Bell Telephone
(513) 565-1514
Fax: (513) 565-7133
llee@cinbell.com

Randy Lincoln
Tennessee One Call System, Inc
(615) 367-1110
Fax: (615) 366-5021
rlincoln@tnonecall.com

Lance D. Loschen
City of Lincoln
(402) 442-8340
Fax: (402) 441-8735
lloschen@ci.lincoln.ne.us

Doug MacMillan
Oklahoma One Call System, Inc.
(405) 840-9955
Fax: (405) 840-9685
dmacmillan@callokie.com

Joe Maniscalco
Lafayette LA Consolidated Govt
(337) 261-8509
Fax: (337) 291-5688

James D. Maniscalco
Utility Notification Center of CO
(303) 232-1991
Fax: (303) 234-1712
jdman@uncc.org

Joe Maricchiolo
Melbourne One Call Services, Inc
61-03 9466 9633
61- 03 9466 9310

Amy Marino
North Carolina One Call
(336) 855-5760
Fax: (336) 299-1914

Jeffrey D. Martens
Diggers Hotline, Inc.
(414) 259-0676, ext 203
Fax: (414) 259-1453
jmartens@diggershotline.com

Shawn Martin
Diggers Hotline of Nebraska
(402) 731-5858
Fax: (402) 731-8656

Charles Mathison
Snohomish County WA
(425) 388-6450
Fax: (425) 388-6449

Pamela M. Mayben
Bonner/Boundry One-Call
(800) 428-4950
Fax: (208) 769-2805
mmayben@guest.net

Diana L. McColgan
Arlington County VA
(703) 228-7137
Fax: (703) 228-3606
dmvvol@co.arlington.va.us

Dennis L. McCormick
City of Cumberland
(301) 759-6626
Fax: (301) 759-6632

Michael K. McDonald
STS, Inc
(602) 242-9294
Fax: (602) 242-9234
mmcdonald@sts-locates.com

William S. McHenry
Dixie Lawn Service
(704) 866-8723
Fax: (704) 866-4714
billmchenry@yahoo.com

Tim McKeown
Florida Power and Light
(904) 254-2376
Fax: (904) 254-2377

Michael T. McNamara
One Call Systems, Inc
(281) 970-0505
Fax: (281) 970-0110

Martin T. Mead
Utility Notification Center of CO
(303) 205-6301
Fax: (303) 234-1712

James T. Mealey, Jr.
City of Nashua
(603) 589-3135
Fax: (603) 589-3169
MealeyJ@ci.nashua.nh.us

Michael Meyer
Diggers Hotline
(414) 259-0676
Fax: (414) 259-1453
mmeyer@diggershotline.com

Stephen R. Meyer
City of Aurora
(630) 844-2800
Fax: (630) 844-2803

Robert Milliken
Underground Services, Inc
(610) 738-8762 x104
Fax: (610) 696-7864
bmil@softdig.com

Kim C. Moody
Izoic, Inc
(602) 953-8854
425-671-2708
kmoody@izoic.com

Becky L. Morris
Utilities Protection
(770) 623-4332
Fax: (770) 623-1847

Scott A. Morrow
Pennsylvania One Call System, Inc
(215) 884-2228
Fax: (412) 464-7104

Raymond Muhs
Julie, Inc

Patricia Mullin
Regional Municipality of Peel
(905) 791-7800x4396
Fax: (905) 791-6728

Truman Murray
City of Jackson
(901) 422-7210
Fax: (901) 422-7221
tmurray@jaxenergy.com

Mahmoud R. Najda
Seminole County Florida
(407) 665-7331
Fax: (407) 665-7456
mnadja@co.seminole.fl.us

Larry D. Nelson
City of Madison
(608) 267-4227
Fax: (608) 264-9275
lnelson@ci.madison.wi.us

James Nics
Utilities Underground Location Ctr
(425) 454-6888
Fax: (425) 451-2385

John Novak
Ohio Utilities Protection Service
(800) 362-2764
Fax: (330) 759-2745

Gregory A. Obsincs
Ohio Utilities Protection Service
(330) 545-7244
Fax: (330) 759-8749
oups@worldnet.att.net

Dan Oglesbee
Mississippi One Call System, Inc
(601) 362-4322
Fax: (601) 366-7666

Joleen O'Harra
Digline
(208) 342-1585
Fax: (208) 429-0224
oharra@digline.com

Ron M. Olitsky
Underground Service Alert
(909) 808-8111
Fax: (909) 808-8101
ron@digalert.org

Kristen E. Ouwerkerk
Arizona Blue Stake, Inc
(602) 659-7500 x 2224
Fax: (602) 659-7510
kouwerkerk@azbsinc.com

David H. Parham
Florida Gas Transmission
(407) 838-7119
Fax: (407) 838-7101

Charles R. Parrish
Las Vegas Valley Water Dist
(702) 870-2011
Fax: (702) 258-3851

Dick Paulk
New Mexico One Call
(505) 241-3407
Fax: (505) 241-3466
rpaulk@pnm.com

Ronald L. Payton
Utility Notification Center of CO
(303) 205-6301
Fax: (303) 234-1712

Max Pendergrass
Kansas One Call System, Inc
(316) 618-8785
Fax: (316) 618-8786
kos@worldnet.att.net

Michael C. Pfender
City of Topeka
(785) 368-3916
Fax: (785) 368-3881
mpfender@topeka.org

Linda A. Phillips
Dig Line, Inc.
(208) 342-1585
Fax: (208) 429-0224
lphillips@digline.com

Okie Picco
Julie, Inc
(815) 741-5000
Fax: (815) 741-5958

Jerri Pierce
Mississippi One Call System, Inc
(601) 362-4322
Fax: (601) 366-7666

Don E. Pinkston, Jr.
GBA Master Series, Inc
(816) 363-2900
Fax: (816) 363-8444
dpinkston@gbams.com

Frank Planton
One-Call Concepts, Inc
(503) 232-1987
Fax: (503) 293-0826

Bob Pose
Diggers Hotline of Nebraska
(515) 987-2800
Fax: (515) 987-3488

David R. Price
Virginia Utility Protection Service
(540) 985-9355 x2006
Fax: (540) 342-8250
dprice@vups.org

Kathy Quartermaine
Tennessee One Call System, Inc
(615) 367-1110
Fax: (615) 366-5021

Annette R. Reburn
Alabama One Call
(205) 414-1840
Fax: (205) 414-1899
areburn@allcall.com

Jed Reed
Utility Notification Center of CO
(303) 205-6301
Fax: (303) 234-1712

Kirk Reh
Third Millenium Corp.
(913) 221-6574
Fax: (913) 888-6168
kirk.reh@3mtcorp.com

Steve Reiswig
Pacific Gas & Electric Co
(925) 866-5306
Fax: (925) 866-3188
spr4@pge.com

Robert B. Rider
City of Cumberland
(301) 759-6623
Fax: (301) 759-6633

Terry Ross
City of Calgary
(403) 268-1247
Fax: (403) 268-1271

Rick Routh
Tennessee One Call System, Inc
(615) 367-1110
Fax: (615) 366-5021
rrouth@tnonecall.com

Kelly Ruddo
One Call Concepts, Inc
(713) 432-0365
Fax: (713) 432-0041
mcruddo@earthlink.net

Joe R. Sacco
Diggers Hotline of Nebraska
(402) 778-3024

Stephanie Saintsing
North Carolina One Call
(336) 855-5760
Fax: (336) 299-1914

Larry J. Schall, P.E.
Shafer Kline & Warren
(816) 756-0444
Fax: (816) 756-1763
laronekamo@aol.com

Connie Schmidt
One Call Concepts, Inc
(414) 259-0676
Fax: (800) 338-3860

Stacy Schneider
Alberta One Call Corp
(403) 531-3712
Fax: (403) 531-3703

Barbara Schuett
One Call Systems, Inc
(718) 631-6700
Fax: (718) 631-6723
bschuett@1-call.com

Patsy Sharpe
Palmetto Utility Protection
(803) 939-1117
Fax: (800) 443-5401
patsy.sharpe@sc1pups.org

John L. Shelton
Shelton Enterprises, Inc
(954) 581-6280
Fax: (954) 583-5800

Philo S. Shelton, III
City of Black Hawk
(303) 582-1324
Fax: (303) 582-2250
pshelton@cityofblackhawk.org

Dan Sherer
Utility Notification Center of CO
(303) 688-3100
Fax: (303) 688-7431
sherer@intermountain-rea.com

Thomas M. Shimon
Kansas One-Call System, Inc
(316) 687-0494
Fax: (316) 687-0629
tshimon@swbell.net

Vaughn Shosted
Questar Gas
(801) 539-3371
Fax: (801) 539-3323
vaughns@mslnpo.qstr.com

Bob Sidhu
New Mexico One Call
(505) 260-1165
Fax: (505) 260-0968

Gary R. Sloman
New Mexico One Call
(505) 260-1165
Fax: (505) 260-0968
gsloman@nmonecall.org

James Glyn Smith
Palmetto Utility Protection
(803) 939-1117
Fax: (803) 939-9296
jglyn.smith@sc1pups.org

Thomas E. Stewart
Oil & Gas Prod Underground Protect Services
(740) 587-0444
Fax: (740) 587-0446
tstewart@ooga.org

Rich Stockand
Pacific Gas & Electric Co
(415) 973-0500
Fax: (415) 973-7707
rmsf@pge.com

Bob Suits
MISS DIG System, Inc
(248) 874-3401
Fax: (248) 874-3410

Mark Sweet
Sunshine State One-Call of Florida, Inc.
(386) 575-2001
Fax: (386) 575-2032
msweet@callsunshine.com

Faisal Syed, P.E., PTOE
Walter P Moore & Assoc, Inc
(214) 740-6252
Fax: (214) 740-6300
Fsyed@walterpmoore.com

Dennis Tanona
MISS DIG System, Inc
(248) 874-3401
Fax: (248) 874-3410

Stephen Tatrai
Pennsylvania One Call System, Inc
(412) 464-7113
Fax: (412) 464-7165

Noel C. Thompson
City of Louisville
(502) 574-3377
Fax: (502) 574-3849
nthompson@louky.org

Edward Trottier
City of Langley
(604) 514-2911
Fax: (604) 530-0199
ed@city.langley.bc.ca

William B. Turner, Jr.
Tennessee One Call System, Inc
(615) 367-1110
Fax: (615) 366-5021
bturner@tnonecall.com

Jim Tyler
Kansas One Call System, Inc
(785) 575-6099

Casey Vann
Vannguard Utility Partners, Inc
(908) 223-2014
vannguardcv@netseape.net

Susan Volkman
One Call Concepts, Inc
(651) 430-1206
Fax: (651) 430-8016

Leigh Walls
Arkansas One-Call Center
(501) 328-2500
Fax: (800) 482-7997

Craig D. Walter
Iowa One Call
(515) 278-8700
Fax: (515) 278-0245
walter1@cnetins.net

Donald M. Ward
One Call Board of Texas
(512) 477-2255
Fax: (512) 467-6664
tassocttq@aol.com

Neil Watts
City of Renton
(425) 430-7278

Joseph Wertz
City of Pearland
(281) 652-1734
Fax: (281) 652-1706
jwertz@pearland.ci.tx.us

Win S. Westfall
Willdan Associates
(916) 924-7000
Fax: (916) 924-3644
wwestfall@willdan.com

G Wight
Wight Consulting Engrs, Inc
(847) 381-1800
Fax: (847) 381-1875
gwight@aol.com

Dave Wiles
CT Consultants
(440) 530-2270
Fax: (440) 951-7487
dwilesectconsultants.com

Dale L. Wilson
Diggers Hotline
(414) 259-0676
Fax: (414) 259-1453

Melinda A. Winney
Southern Nevada Environmental
(702) 248-5370
Fax: (702) 248-8036
melindawinney@snei.com

Stephen H. Wright
Dig Safely New York, Inc
(716) 724-8469
Fax: (716) 724-8775

John Zizolfo
One Call Users' Council, Inc
(212) 330-4235
Fax: (212) 447-6913
zizolfoj@coned.com

Ben Zweifel
Diggers Hotline
(414) 259-0676
Fax: (414) 259-1453
bzweifel@diggershotline.com

One-Call Systems international (OCSI) Directory

ALABAMA
1-800-292-8525

ALABAMA ONE-CALL
Center # 205-252-4444
Alternate Center # 205-414-1840 (Administrative)
PO Box 1476
Birmingham, AL 35201
Contact: Annette R. Reburn, Executive Director
Contact # 205-414-1840
Fax # 205-414-1899
Do you accept FAX notifications? Yes
Type Center: Nonprofit
Staff Coverage Area: Statewide
Members: 340
Hours of Operation: 0700-1700 M - F
Legislation: Yes, 94-487
Mandatory Membership: No
Notice: 2 working days - notice good for 14 calendar days
Handle Emergency Notice?: Yes
E-mail: areburn@al1call.com
Web Site: www.al1call.com
www.allcall.com

ALASKA
1-800-478-3121

ALASKA DIGLINE, INC.
Locate Call Center of Alaska
Center # 907-278-3121
Alternate Center #
2221 E. Northern Lights Blvd., Suite 136
Anchorage, AK 99508
Contact: Louise Nelson, Manager
Contact # 907-279-1122
Fax # 907-278-0696
Do you accept FAX notifications? Yes; 907-278-0696
Type Center: Contract
Coverage Area: Statewide
Members: 26
Hours of Operation: 0800-1700 (M-F)
Legislation: Yes
Mandatory Membership: No
Notice: 2 days
Handle Emergency Notice? Yes
E-mail: Lcc@akonecall.com
Web Site: www.akonecall.com

ARIZONA
1-800-STAKE-IT (1-800-782-5348)

ARIZONA BLUE STAKE
Center # 602-263-1100 (inside Maricopa County)
1-800-STAKE-IT (1-800-782-5348 outside Maricopa County)
Alternate Center # 602-659-7500
4415 S. Wendler Dr., Suite 105
Tempe, AZ 85282
Contact: Sandra M. Holmes, Executive Director
Contact # 602-659-7500 x. 2223
Fax # 602-659-7510
Do you accept FAX notifications? Yes; 602-659-7510
Type Center: Nonprofit Staff
Coverage Area: Statewide
Members: 698
Hours of Operation: 0600-1630
Legislation: Yes
Mandatory Membership: Yes
Notice: 2 working days, excluding weekends and state-recognized holidays, marks valid for 15 working days excluding weekends and state-recognized holidays
Handle Emergency Notice? Yes
E-mail: smholmes@azbsinc.com
Web Site: www.azbsinc.com

ARKANSAS
1-800-482-8998

ARKANSAS ONE CALL SYSTEM, INC.
Center # 501-328-2500
Alternate Center # 501-328-2525
505 Industrial Blvd.
Conway, AR 72032
Contact: Dale Enoch, Executive Director
Contact # 501-328-2500
Fax # 501-328-2522
Do you accept FAX notifications? Yes (with approval) Fax: 800-482-7997
Type Center: In-House
Coverage Area: Statewide
Members: 675
Hours of Operation: 24 hours/7 days
Legislation: Yes
Mandatory Membership: Yes, partial
Notice: 2-10 working days
Handle Emergency Notice? Yes
E-mail: denoch@arkonecall.com
Web Site: www.arkonecall.com

CALIFORNIA
1-800-227-2600

UNDERGROUND SERVICE ALERT NORTH
Center # 1-800-227-2600
Alternate Center #
4090 Nelson Avenue - Suite A
Concord, CA 94520-1232
Contact: Mike Heyer, General Manager
Contact # 925-798-9504
Fax # 925-798-9506
Do you accept FAX notifications? Yes
Type Center: Nonprofit Staff
Coverage Area: Central and Northern CA
Members: 1,062
Hours of Operation: 0600-1900 (PST) M-F
Legislation: Yes
Mandatory Membership: Yes
Notice: 2 working days, but not more than 14 calendar days
Handle Emergency Notice? No
E-mail: usanorth@usan.org
Web Site: usanorth.org

UNDERGROUND SERVICE ALERT OF SOUTHERN CALIFORNIA
Center # 909-808-8100
Alternate Center # 1-800-422-4133
1760 California Ave., #101
Corona, CA 92881
Contact: Ron Olitsky, President
Contact # 909-808-8111
Fax # 909-808-8101
Do you accept FAX notifications? No
Type Center: Nonprofit Staff
Coverage Area: Southern California
Members: 829
Hours of Operation: 0600-1900 M - F
Legislation: Yes
Mandatory Membership: Yes
Notice: 2 working days but not more than 14 calendar days

Handle Emergency Notice? Yes
E-mail: ron@digalert.org
Web Site: www.digalert.org

COLORADO

1-800-922-1987

UTILITY NOTIFICATION CENTER OF COLORADO
Center # 1-800-922-1987
Administration # 1-800-833-9417
12600 W. Colfax Avenue, Suite B310
Lakewood, CO 80215
Contact: J.D. Maniscalco, Executive Director
Contact # 303-232-1991
Fax # 303-234-1712
Do you accept FAX notifications? Yes; 303-205-6335
Type Center: Nonprofit Staff
Coverage Area: Statewide
Members: 866
Hours of Operation: 0700-1700 M-F; Emergency 24 hour coverage, 7 days a week
Legislation: Yes
Mandatory Membership: Yes
Notice: 2 days not to include the day of notice
Handle Emergency Notice? Yes
E-mail: jdman@@uncc.org
Web Site: http://www.uncc.org

CONNECTICUT

1-800-922-4455

CALL BEFORE YOU DIG
Center # 1-800-922-4455
Alternate Center # 203-281-5435
2040 Whitney Avenue
Hamden, CT 06517
Contact: William Petersen
Contact # 203-248-5502
Fax # 203-248-6448
E-mail: bpetersen@cbyd.com
Do you accept FAX notifications? Yes; 203-248-6448
Type Center: Contract
Coverage Area: Statewide
Members: 316
Hours of Operation: 24 hours - 7 days a week
Legislation: Yes
Mandatory Membership: Yes
Notice: Minimum of 2 working days in advance, no more than 30 days in advance
Handle Emergency Notice? Yes
Web Site: www. cbyd .com OR www.cbyd.org OR www.cbyd.net OR www.cbyd.cc

DELAWARE

1-800-282-8555

MISS UTILITY OF DELMARVA
Delaware, Eastern Virginia, Eastern Shore Maryland
Center # 1-800-282-8555
Alternate Center # 1-800-441-8355
7223 Parkway Drive
Hanover, MD 21076
Contact: Jim Holzer, Director of Operations
Contact # 410-712-0056
Fax # 410-712-0062
Do you accept FAX notifications? Yes 410-712-0062
Type Center: Contract
Coverage Area: Statewide
Members: 77
Hours of Operation: 24 hrs.
Legislation: Yes
Mandatory Membership: Yes
Notice: 48 hours required
Handle Emergency Notice? Yes

FLORIDA

1-800-432-4770

SUNSHINE STATE ONE-CALL OF FLORIDA, INC.
Center # 1-800-432-4770
Alternate Center # 386-575-2025
11 Plantation Road
DeBary, FL 32713
Contact: Mark Sweet, Executive Director
Contact # 386-575-2001
Fax # 386-575-2032 (primary) 386-575-2030 (secondary)
Do you accept FAX notifications? Yes; Not Published
Type Center: Not-for-profit
Coverage Area: Statewide
Members: 667
Hours of Operation: 0700-1700 M - F
Legislation: Yes
Mandatory Membership: except small cities
Notice: 2-5 business days
Handle Emergency Notice? Yes
E-mail: msweet@callsunshine.com
Web Site: www.callsunshine.com

GEORGIA

1-800-282-7411

UTILITIES PROTECTION CENTER, INC.
Center # 1-800-282-7411
Alternate Center # 770-623-4344
3400 Summit Ridge Parkway
Duluth, GA 30096
Contact: Claudette L. Campbell, Executive Director
Contact # 770-623-4332, ext. 200
Fax # 770-623-1847
Do you accept FAX notifications? No
Do you accept Web notifications? Yes
Type Center: Nonprofit Staff
Coverage Area: Statewide
Members: 450+
Hours of Operation: 0700-1800
Legislation: Yes
Mandatory Membership: Yes
Notice: 48 hours, excluding the day called in
Handle Emergency Notice? Yes
E-mail: ccampbell@gaupc.com
Web Site: www.gaupc.com

HAWAII

1-800-227-2600

UNDERGROUND SERVICE ALERT NORTH
Center # 1-800-227-2600
Alternate Center #
4090 Nelson Avenue - Suite A
Concord, CA 94520-1232
Contact: Mike Heyer, General Manager
Contact # 925-798-9504
Fax # 925-798-9506
Do you accept FAX notifications? Yes
Type Center: Nonprofit Staff
Coverage Area: Entire State
Members: 9
Hours of Operation: 0600-1900 (PST) M-F
Legislation: No (being prepared)
Mandatory Membership: No
Notice: 2 working days, but not more

than 14 calendar days
Handle Emergency Notice? No
E-mail: usanorth@usan.org
Web Site: usanorth.org

IDAHO

DIG LINE, INC.
Center # 1-800-342-1585
Alternate Center # (208) 342-1585
3350 Americana Terrace, Suite 370
Boise, ID 83706
Contact: Linda Phillips, Manager
Contact # (208) 342-1585
Fax # 208-429-0224
Do you accept FAX notifications?
Yes; 1-800-342-1586
Type Center: Contract
Coverage Area: Idaho counties: Ada, Adams, Bannock, Bear Lake, Bingham, Blaine, Boise, Bonneville, Butte, Camas, Canyon, Caribou, Cassia, Clark, Custer, Elmore, Franklin, Fremont, Gem, Gooding, Idaho, Jefferson, Jerome, Lemhi, Lincoln, Madison, Minidoka, Oneida, Owyhee, Payette, Power, Teton, Twin Falls, Valley, Washington, Nez Perce, Lewis
Members: 150
Hours of Operation: 24 hours
Legislation: Yes
Mandatory Membership: Yes
Notice: 2 business days
Handle Emergency Notice? Yes
E-mail: lphillips@digline.com
E-mail 2: digline@digline.com
Web Site: digline@digline.com

ONE-CALL CONCEPTS - IDAHO
Center # 1-800-626-4950
Alternate Center # 1-800-822-1974
1415 SE Ankeny
Portland, OR 97214
Contact: Frank Planton, General Manager
Contact # 503-232-1987
Fax # 503-293-0826
Do you accept FAX notifications?
Yes; 877-695-2466
Type Center: Contract
Coverage Area: 1-800-626-4950: Bonner and Boundary Counties; Washington, Pend Oreille County; Montana, Lincoln and Sanders Counties 1-800-822-1974;
Members: 50
Hours of Operation: 24 hrs./7 days, all locates
Legislation: Yes
Mandatory Membership: Yes
Notice: 2 working days
Handle Emergency Notice? Yes

PALOUSE EMPIRE UNDERGROUND COORDINATING COUNCIL
Center # 1-800-822-1974
Alternate Center #
P.O. Box 72
Pullmann, WA 99163
Contact: Alan Anderson, President
Contact # 509-334-4555
Fax # 509-332-8487
Do you accept FAX notifications? No
Type Center: Nonprofit Staff
Coverage Area: Northern Idaho-Clearwater and Latah Counties
Members: 40
Hours of Operation: 0730-1700
Legislation: Yes
Mandatory Membership: No
Notice: 2 days
Handle Emergency Notice? Yes

BONNER/BOUNDRY ONE-CALL
Center # 1-800-626-4950
Alternate Center #
1919 N. 3rd Street
Coeur D'Alene, ID 83814
Contact: P. Marie Mayben, Manager
Contact # 1-800-428-4950
Fax # 208-769-2805
Do you accept FAX notifications?
Yes; 208-664-2346
Type Center: Contract
Coverage Area: Bonner and Boundry Counties, Idaho
Members: 50
Hours of Operation: 24 hours
Legislation: Yes
Mandatory Membership: Yes
Notice: 48 hours
Handle Emergency Notice? Yes

KOOTENAI COUNTY ONE-CALL
Center # 1-800-428-4950
Alternate Center #
1919 N. 3rd Street
Coeur D'Alene, ID 83814
Contact: P. Marie Mayben, Manager
Contact # 1-800-428-4950
Fax # 208-769-2805
Do you accept FAX notifications?
Yes; 208-664-2346
Type Center: Contract
Coverage Area: Kootenai County
Members: 100
Hours of Operation: 24 hours
Legislation: Yes
Mandatory Membership: Yes
Notice: 48 hours
Handle Emergency Notice? Yes

SHOSHONE - BENEWAH ONE-CALL
Center # 1-800-398-3285
Alternate Center #
1919 N. 3rd Street
Coeur D'Alene, ID 83814
Contact: P. Marie Mayben, Manager
Contact # 1-800-398-3285
Fax # 208-769-2805
Do you accept FAX notifications?
Yes; 208-664-2346
Type Center: Contract
Coverage Area: Shoshone and Benewah Counties
Members: 50
Hours of Operation: 24 hours
Legislation: Yes
Mandatory Membership: Yes
Notice: 48 hours
Handle Emergency Notice? Yes

UTILITIES UNDERGROUND LOCATION CENTER
Center # 1-800-424-5555
Alternate Center #
13427 NE 20th St., #120
Bellevue, WA 98005
Contact: Tom Odegaard, Manager
Contact # 425-454-6888
Fax # 425-451-2385
Do you accept FAX notifications? Yes
800-461-7670
Type Center: Contract
Coverage Area: Clearwater, Idaho, Lewis and Nez Perce Counties
Members: 13
Hours of Operation: 0600-1700
M - F 24 hour emergency service
Legislation: Yes
Mandatory Membership: No
Notice: 2 days
Handle Emergency Notice? Yes

ILLINOIS
1-800-892-0123

Joint Utility Locating Information for Excavators (JULIE, Inc.)
Center # 815-741-5000
3275 Executive Drive
Joliet, IL 60431
Contact: Mark A. Frost, Executive Director
Contact # 815-741-5000
Fax # 815-741-5958
Do you accept FAX notifications? Yes; 800-292-0222
Type Center: Member Utility
Coverage Area: Statewide (except City of Chicago)
Members: 1,055
Hours of Operation: 24 hours
Legislation: Yes
Mandatory Membership: Yes
Notice: Illinois state law requires at least a 48-hour notice (two working days), but no more than a 14-calendar day advance notice prior to the start of excavation.
Handle Emergency Notice? Yes
E-mail Contact: (Kevin Chmura, Public Relations Manager) chmura@julie1call.com Web Site: www.julie1call.com OR www.julieonecall.com

DIGGER (CHICAGO UTILITY ALERT NETWORK)
Center # 312-744-7000
Alternate Center #
30 N. LaSalle Street - Suite 1101
Chicago, IL 60602
Contact: Monty Camano
Contact # 312-744-4062
Fax # 312-742-0950
Do you accept FAX notifications? Yes - Emergency Only
Type Center: Nonprofit Staff
Coverage Area: Chicago Area
Members: 17
Hours of Operation: 24 hours/365 days
Legislation: Yes
Mandatory Membership: Yes
Notice: 48 hours
Handle Emergency Notice? Yes

INDIANA
1-800-382-5544

INDIANA UNDERGROUND PLANT PROTECTION SERVICE
Center # 877-230-0495
Alternate Center # 317-893-1400
P. O. Box 219
Greenwood, IN 46143
Contact: Millie Marsh, Executive Director
Contact # 877-230-0495
Fax # 877-230-0496
Do you accept FAX notifications? Yes; 877-230-0496 (Toll Free)
Type Center: In House
Coverage Area: Statewide
Members: 347
Hours of Operation: 24 hours/365 days
Legislation: Yes
Mandatory Membership: No
Notice: 2 full working days
Handle Emergency Notice? Yes
E-mail: mmarsh@iupps.org
Web Site: www.iupps.org

IOWA
1-800-292-8989

IOWA ONE-CALL (Underground Plant Location Service, Inc.)
Center # 1-800-292-8989
Center Manager # 563-322-2400
320 Le Claire Street
Davenport, IA 52801
Contact: Craig D. Walter, Administrative Director
Contact # 515-278-8700
Fax # 515-278-0245
Do you accept FAX notifications? Yes
Type Center: Contract
Coverage Area: Statewide
Members: 1,400
Hours of Operation: 24 hours/7 days
Legislation: Yes
Mandatory Membership: Yes
Notice: 2 working days
Handle Emergency Notice? Yes
E-mail: onecall@iowaonecall.com
Web Site: www.iowaonecall.com

KANSAS
1-800-DIG-SAFE

KANSAS ONE-CALL SYSTEM, INC.
Center # 1-800-DIG-SAFE
Alternate Center # 316-687-2470
Administrative # 316 687 2102
8100 East 22nd Street North - Bldg 2300
Wichita, KS 67226
Contact: Tom Shimon, Executive Director
Contact # 316-687-0494
Fax # 316-687-0629
Do you accept FAX notifications? Yes, 316-687-3753
Type Center: Contract
Coverage Area: Statewide
Members: 609
Hours of Operation: 0700-1800 M - F/ Emergencies - 24hrs.
Legislation: Yes
Mandatory Membership: Yes (except water & sewer)
Notice: 2 business days
Handle Emergency Notice? Yes
E-mail: tshimon@swbell.net

KENTUCKY
1-800-752-6007

KENTUCKY UNDERGROUND PROTECTION INC.
Center # 1-800-752-6007
Alternate Center # 502-266-5677
9322 Taylorsville Road - Suite 4A
Louisville, KY 40299
Contact: Sharon M. Able, Executive Director
Contact # 502-633-0171 or 877-711-6225
Fax # 502-633-0710
Do you accept FAX notifications? Yes, Internet Interactive programs also
Type Center: Self-run
Coverage Area: Statewide
Members: 210
Hours of Operation: 0700-1800 M - F
Legislation: Yes
Mandatory Membership: No
Notice: 2 business days
Handle Emergency Notice?Yes
E-mail: kupi@bellsouth.net
Web Site: www.kyunderground.com

LOUISIANA

1-800-272-3020

LOUISIANA ONE CALL SYSTEM, INC.
Center # 1-800-272-3020
Administrations # 225-275-3700 or 1-800-584-4274
2215 West Boardwalk Drive
Baton Rouge, LA 70816
Contact: David Frey, Manager/ Damage Prevention
Contact # 225-275-3700 Administrations Fax # 225-272-1967
Do you accept FAX notifications? Yes
Type Center: Contract
Coverage Area: Statewide
Members: 812
Member Notification Terminals: 1,182
Hours of Operation: 0700-1800
Legislation: Yes
Mandatory Membership: Yes, in Regional Notification Program
Notice: 48 to 120 hours
Handle Emergency Notice? Yes

MAINE

1-888-DIG-SAFE

DIG SAFE SYSTEM, INC.
Center # 781-721-0990
Alternate Center # 877-344-7233 (877-DIG-SAFE)
331 Montvale Ave.
Woburn, MA 01801
Contact: Robert S. Finelli, Executive Director
Contact # 781-721-1191
Fax # 781-721-0047
Do you accept FAX notifications? Yes; 781-721-0047
Type Center: Not-for-Profit
Coverage Area: Statewide
Members: 380
Hours of Operation: 0600-1800 M-F
Legislation: Yes
Mandatory Membership: Yes
Notice: 72 hours, excluding weekends and holidays
Handle Emergency Notice? Yes
Web Site: www.digsafe.com

MARYLAND

MISS UTILITY
Center # 1-800-257-7777
Alternate Center #
7223 Parkway Drive
Hanover, MD 21076
Contact: Jim Holzer, Director of Operations
Contact # 410-712-0056
Fax # 410-712-0062
Do you accept FAX notifications? Yes 410-712-0062
Type Center: Contract
Coverage Area: Statewide
Members: 225
Hours of Operation: 0700-1700
Legislation: Yes
Mandatory Membership: Yes
Notice: 48 hours required
Handle Emergency Notice? Yes

MISS UTILITY OF DELMARVA
Center # 1-800-282-8555
Alternate Center # 1-800-441-8355
7223 Parkway Drive
Hanover, MD 21076
Contact: Jim Holzer, Director of Operations
Contact # 410-712-0056
Fax # 410-712-0062
Do you accept FAX notifications? Yes 410-712-0062
Type Center: Contract
Coverage Area: Eastern Shore of Maryland
Members: 71
Hours of Operation: 0700-1700
Legislation: Yes
Mandatory Membership: Yes
Notice: 48 hours required
Handle Emergency Notice? Yes

MASSACHUSETTS

1-888-DIG-SAFE

DIG SAFE SYSTEM, INC.
Center # 781-721-0990
Alternate Center # 877-DIG-SAFE (877-344-7233)
331 Montvale Ave.
Woburn, MA 01801
Contact: Robert S. Finelli, Executive Director
Contact # 781-721-1191
Fax # 781-721-0047
Do you accept FAX notifications? Yes; 781-721-0047
Type Center: Not-for-Profit Staff
Coverage Area: Statewide
Members: 380
Hours of Operation: 0600-1800 M-F
Legislation: Yes
Mandatory Membership: Yes
Notice: 72 hours, excluding weekends and holidays
Handle Emergency Notice? Yes
Web Site: www.digsafe.com

MICHIGAN

1-800-482-7171

MISS DIG SYSTEM, INC.
Center # 1-800-482-7171
Alternate Center # 248-647-7344
1030 Featherstone Road
Pontiac, MI 48342-1830
Contact: Kathleen A. Fournier, Executive Director
Contact # 248-874-3401
Fax # 248-874-3410
Do you accept FAX notifications? No
Type Center: Nonprofit
Coverage Area: Statewide
Members: 890
Hours of Operation: 24 hours/7 days
Legislation: Yes
Mandatory Membership: Yes
Notice: 3 full working days
Handle Emergency Notice? Yes
E-mail: kafournier@mindspring.com
E-mail 2: kafournier@missdig.org
Web Site: www.missdig.org

MINNESOTA

1-800-252-1166

GOPHER STATE ONE CALL
Center # 1-800-252-1166
Alternate Center # 651-454-0002
2020 Centre Point Blvd.
Mendota Heights, MN 55120
Contact:Tonya Bethke, General Manager
Contact # 651-454-8388
Fax # 651-454-0170
Do you accept FAX notifications? Yes
Type Center: Contract

Coverage Area: Statewide
Members: 1,400
Hours of Operation: 24 hours
Legislation: Yes
Mandatory Membership: Yes
Notice: 48 hours, excluding weekends and holidays
Handle Emergeny Notice? Yes
E-mail: gsoc@isd.net
Web Site: www.gopherstateonecall.org

MISSISSIPPI
1-800-227-6477

MISSISSIPPI ONE-CALL SYSTEM, INC.
Center # 601-362-4374
Alternate Center # 800-227-6477
5258 Cedar Park Drive - Suite H
Jackson, MS 39206
Contact: Sam Johnson
Contact # 601-362-4322
Fax # 601-366-7666
Do you accept FAX notifications?
Yes; 601-362-7533
Type Center: In-House
Coverage Area: Statewide
Members: 340
Hours of Operation: 24 hours/7 days
Legislation: Yes
Mandatory Membership: No
Notice: 2 working days
Handle Emergency Notice? Yes
E-mail: ms1call@bellsouth.net
Web Site: www.ms1call.org

MISSOURI
1-800-344-7483

MISSOURI ONE-CALL SYSTEM, INC.
Center # 573-635-1550
Alternate Center # 412-415-5058
728 Heisinger Road
Jefferson City, MO 65109-4711
Contact: John Lansford, Executive Director
Contact # 573-635-1818
Fax # 573-635-8402
Do you accept FAX notifications?
Yes; 1-800-217-3717
Type Center: Contract
Coverage Area: Statewide
Members: 270
Hours of Operation: 24 hours/7 days
Legislation: Yes
Mandatory Membership: Yes
Notice: 2 working days
Handle Emergency Notice? Yes
E-mail: JohnL@occinc.com
Web Site: www.mo1call.org

MONTANA

UTILITIES UNDERGROUND LOCATION CENTER
1-800-424-5555
Center # 877-668-4001
Alternate Center # 406-442-3070
140 Wedgewood Lane, Helena, Montana 59601
Contact: A. F. "Bud" Criner
Contact # 406-442-3070
Fax # 406-443-3554
E-mail: uulcrep@msn.com
Do you accept FAX notifications? Yes
Type Center: Contract
Coverage Area: Statewide
Members: 240
Hours of Operation: 24 hours/7 days/ 365 days
Legislation: Yes
Mandatory Membership: yes
Notice: 2 business days
Handle Emergency Notice? Yes
E-mail: through web site
Web Site: www.callbeforeyoudig.com

MONTANA ONE CALL CENTER
Center # 1-800-551-8344
Alternate Center #
13 East Oregon Street
Kalispell, MT 59901
Contact: Doug R. Johns
Contact # 406-752-6811
Fax # 406-752-7306
Do you accept FAX notifications? Yes
406-752-7306
Type Center: Contract
Coverage Area: Northwest Region of State; All of Flathead & Lincoln Counties and the North half of Lake County
Members: 27
Hours of Operation: 24 hours
Legislation: Yes
Mandatory Membership: Yes
Notice: 2 business days
Handle Emergency Notice? Yes
E-mail: mpn@digisys.net
Web Site: www.udig.org

NEBRASKA
1-800-331-5666

DIGGERS HOTLINE OF NEBRASKA
Center # 1-800-331-5666
Alternate Center # 402-330-1866
13205 C Street
Omaha, NE 68144
Contact: Teresa Wilgus, Director
Fax # 402-330-5626
Contact # 402-330-8377
Fax# 402-330-5626
Do you accept FAX notifications?
Yes; 402-330-5626 or 800-896-0664
Type Center: Contract
Coverage Area: Statewide
Members: 612
Hours of Operation: 24 hours/365 days
Legislation: Yes
Mandatory Membership: Yes
Notice: 48 hours excluding weekends and state/federal holidays
Handle Emergency Notice? Yes
E-mail: dhadmin@gpcom.net
E-mail 2: ne-dh@gpcom.com
Web Site: www.ne-diggers.com

NEVADA
1-800-227-2600

UNDERGROUND SERVICE ALERT NORTH
Center # 1-800-227-2600
Alternate Center #
4090 Nelson Avenue - Suite A
Concord, CA 94520-1232
Contact: Mike Heyer, General Manager
Contact # 925-798-9504
Fax # 925-798-9506
Do you accept FAX notifications? Yes
Type Center: Nonprofit Staff
Coverage Area: Statewide
Members: 137
Hours of Operation: 0600-1900 (PST) M-F
Legislation: Yes
Mandatory Membership: Yes
Notice: 2 working days, but not more than 14 calendar days
Handle Emergency Notice? No
E-mail: usanorth@usan.org
Web Site: usanorth.org

NEW HAMPSHIRE

1-888-DIG-SAFE

DIG SAFE SYSTEM, INC.
Center # 781-721-0990
Alternate Center # 877-344-7233 (877-DIG-SAFE)
331 Montvale Ave
Woburn, MA 01801
Contact: Robert S. Finelli, Executive Director
Contact # 781-721-1191
Fax # 781-721-0047
Do you accept FAX notifications? Yes 781-721-0047
Type Center: Not-for-Profit Staff
Coverage Area: Statewide
Members: 380
Hours of Operation: 0600-1800 M-F
Legislation: Y
Mandatory Membership: Y
Notice: 72 hours, excluding weekends and holidays
Handle Emergency Notice? Y
Web Site: www.digsafe.com

NEW JERSEY

1-800-272-1000

NEW JERSEY ONE CALL
Center # 1-800-272-1000
Alternate Center # 732-394-3000
399 Hoes Lane
Piscataway, NJ 08854
Contact: Barbara Bye
Contact # 732-394-3010
Fax # 732-394-3011
Do you accept FAX notifications? Yes
Type Center: Contract w/State
Coverage Area: Statewide
Members: Operators of underground utilities
Hours of Operation: 24 hours
Legislation: Yes
Mandatory Memebership: operators of underground utilities
Notice: 3 full business days
Handle Emergency Notice? Yes
E-mail: bbye@i-2000.com
Web Site: www.nj1-call.org

NEW MEXICO

1-800-321-ALERT (1-800-321-2537)

NEW MEXICO ONE CALL SYSTEM, INC.
Center # 505-260-1990 (local)
Alternate Center # 800-321-2537
PO Box 27192
Albuquerque, NM 87125-7192
Contact: Harry Sherlock, General Manager
Contact # 505-260-1165
Fax # 505-260-0968
Do you accept FAX notifications? Yes; 505-260-0968 and 1-800-260-0950
Type Center: Nonprofit Staff
Coverage Area: Statewide
Members: 151
Hours of Operation: 0700-1700 M-F
Legislation: Yes
Mandatory Membership: Pipelines (including gas distribution) in or near public right-of-way - Yes Others: No
Notice: 2 working days
Handle Emergency Notice? Yes
E-mail: sherlock@nmonecall.org
Web Site: www.nmonecall.org

LAS CRUCES-DONA ANA BLUE STAKES
Center # 505-526-0400 or 1-888-526-0400
Alternate Center # 505-526-0500 After hours
680 N. Motel Blvd.
Las Cruces, NM 88005
Contact: Mary Alba
Contact # 505-526-0400
Fax # 505-526-6022
Do you accept FAX notifications? Yes; 505-526-6022
Type Center: Self Run
Coverage Area: Dona Ana County
Members: 7
Hours of Operation: 0700-1600 (24 hrs - emergencies)
Legislation: Yes
Mandatory Membership: No
Notice: 48 hours (2 working days in advance)
Handle Emergency Notice? Yes (1 hour response time)

NEW YORK

DIG SAFELY NEW YORK (Formerly UFPO)
Center # 1-800-962-7962
Alternate Center # 315-437-7333
5063 Brittonfield Parkway
East Syracuse, NY 13057
Contact: Robert Foster, Executive Director
Contact # 315-437-7394
Fax # 315-437-2621
Do you accept FAX notifications? No (Use i-notice via Internet)
Type Center: Not-for-Profit - In-house
Coverage Area: All of New York State except New York City and Long Island
Members: 1,287
Hours of Operation: 24 hours/7 days
Legislation: Yes
Mandatory Membership: Yes
Notice: 2 – 10 days, excluding day of call
Handle Emergency Notice? Yes
E-mail: rfoster@digsafelynewyork.com
Web Site: www.digsafelynewyork.com

NEW YORK CITY - LONG ISLAND ONE CALL CENTER
Center # 1-800-272-4480
Alternate Center # 718-631-6700
36-35 Bell Blvd., Suite 202
Bayside, NY 11361
Contact: Barbara Schuett, Director
Contact # 718-631-6700
Fax # 718-631-6723
Do you accept FAX notifications? No
Type Center: Contract
Coverage Area: New York City and Long Island areas of New York
Members: 88
Hours of Operation: 0700-1700 M-F
Legislation: Yes
Mandatory Membership: Yes
Notice: 2-10 days, excluding day of call
Handle Emergency Notice? Yes
E-mail: bschuett@1-call.com
E-mail 2: nycocuc@aol.com

NORTH CAROLINA
1-800-632-4949

THE NORTH CAROLINA ONE-CALL CENTER, INC.
Center # 1-800-632-4949
Alternate Center # 336-855-5760
2300 West Meadowview Road - Suite 227
Greensboro, NC 27407
Contact: George Glenn, Executive Director
Contact # 336-855-5760
Fax # 336-299-1914
Do you accept FAX notifications? Yes; 336-299-1914
Type Center: Nonprofit
Coverage Area: Statewide
Members: 241
Hours of Operation: 0700-1900
Legislation: Yes
Mandatory Membership: No
Notice: 48 hours good to 15 working days
Handle Emergency Notice? Yes
E-mail: george@ncocc.org
Web Site: www.ncocc.org

NORTH DAKOTA
1-800-795-0555

NORTH DAKOTA ONE-CALL
Center # 1-800-795-0555
Alternate Center #
2020 Center Point Blvd.
Mendota Heights, MN 55120
Contact: Tonya Bethke, General Manager
Contact # 800-422-1242
Fax # 877-651-7030
Do you accept FAX notifications? Yes
Type Center: Contract
Coverage Area: Statewide
Members: 500
Hours of Operation: 24 hrs.
Legislation: Yes
Mandatory Membership: Yes
Notice: 48 hours, excluding weekends and holidays
Handle Emergency Notice? Yes
E-mail: tonyabethke@isd.net

OHIO
1-800-362-2764

OHIO UTILITIES PROTECTION SERVICE
Center # 1-800-362-2764
Alternate Center #
4740 Belmont Avenue
Youngstown, OH 44505-1014
Contact: Gregory A. Obsincs, President
Contact # 330-545-7244
Fax # 330-759-8749
Do you accept FAX notifications? No
Type Center: Nonprofit Staff
Coverage Area: Statewide
Members: 890
Hours of Operation: 24 hours/7 days
Legislation: Yes
Mandatory Membership: Yes
Notice: 48 hours
Handle Emergency Notice? Yes
E-mail: oups@worldnet.att.net
Web Site: www.oups.org

OIL AND GAS PRODUCERS UNDERGROUND PROTECTION SERVICE
Center # 1-800-925-0988
Alternate Center # 740-587-0444
P. O. Box 535
Granville, OH 43023
Contact: Stacey Wright
Contact # 740-587-0444
Fax # 740-587-0446
Do you accept FAX notifications? Yes; 740-587-0446
Type Center: Crude Oil and Natural Gas Producer Underground Pipeline Facilities
Coverage Area: Statewide
Members: 181
Hours of Operation: 0800 - 1700, M-F
Legislation: Yes
Mandatory Membership: Yes
Notice: 48 hours
Handle Emergency Notice? No
E-mail: ooga@nextek.net
Web Site: www.ooga.org OR www.ogpups.com

OKLAHOMA
1-800-522-6543

CALL OKIE (OKLAHOMA ONE-CALL SYSTEMS, INC.)
Center # 1-800-522-6543
Alternate Center # 1-800-654-8249
2831 N.W. 59th Street
Oklahoma City, OK 73112-7005
Contact: James A. Hill, Executive Director
Contact # 405-840-9955
Fax # 405-840-9685
Do you accept FAX notifications? Yes; 800-377-1339
Type Center: Nonprofit staff
Coverage Area: Statewide
Members: 710
Hours of Operation: 0700-1800
Legislation: Yes
Mandatory Membership: No
Notice: 48 hours
Handle Emergency Notice? Yes
E-mail: jhill@callokie.com
Web Site: www.callokie.com

OREGON
1-800-332-2344

OREGON UTILITY NOTIFICATION CENTER
Center # 1-800-332-2344
Alternate Center # 503-246-6699
1415 S.E. Ankeny
Portland, OR 97214
Contact: Frank Planton, General Manger
Contact # 503-232-1987
Fax # 503-293-0826
Do you accept FAX notifications? Yes; 503-293-0826
Type Center: Contract
Coverage Area: Statewide
Members: 906
Hours of Operation: 24 hours/7 days
Legislation: Yes
Mandatory Membership: Yes
Notice: 2 days-life of the project. Marks maintained by Excavator.
Handle Emergency Notice? Yes

PENNSYLVANIA

1-800-242-1776

PENNSYLVANIA ONE CALL SYSTEM, INC.
Center # 1-800-242-1776
Alternate Center # 412-464-7100
925 Irwin Run Road
West Mifflin, PA 15122
Contact: William G. Kiger,
Executive Director
Contact # 412-464-7111
Fax # 412-464-7104
Do you accept FAX notifications?
Yes; 412-464-7135
Type Center: Nonprofit
Coverage Area: Statewide
Members: 2,422
Hours of Operation: 24 hours/7 days
Legislation: Yes
Mandatory Membership: Yes Response to all notices mandatory through the Center Automated System "KARL" at 800-222-6470
Notice: Construction phase - 3 to 10 working days; Design phase - 10 to 90 working days Handle Emergency Notice? Yes
E-mail: bgkiger@pa1call.org
Web Site: www.paonecall.org
Remarks: Annual Service Fee of $50 is charged to excavators and is required by law. The fee is used to offset the cost of Municipal participation.

RHODE ISLAND

1-888-DIG-SAFE

DIG SAFE SYSTEM, INC.
Center # 781-721-0990
Alternate Center # 877-DIG-SAFE
(877-344-7233)
331 Montvale Ave.
Woburn, MA 01801
Contact: Robert S. Finelli,
Executive Director
Contact # 781-721-1191
Fax # 781-721-0047
Do you accept FAX notifications?
Yes; 781-721-0047
Type Center: Not-for-Profit
Coverage Area: Statewide
Members: 380
Hours of Operation: 0600-1800 M-F
Legislation: Yes
Mandatory Membership: Yes
Notice: 48 hours, excluding weekends and holidays
Handle Emergency Notice? Yes
Web Site: www.digsafe.com

SOUTH CAROLINA

1-888-SC1-PUPS (1-888-721-7877)

PALMETTO UTILITY PROTECTION SERVICE INC. "PUPS"
Center # 1-888-SC1-PUPS
(1-888-721-7877)
Alternate Center # 800-922-0983
1219 Knox Abbott Drive, Suite 100
Cayce, SC 29033
Contact: James Glyn Smith,
Executive Director
Contact # 803-939-1117
Fax # 1-803-939-0704
Do you accept FAX notifications? Yes
Type Center: Member Utility
Coverage Area: Statewide
Members: 182
Hours of Operation: Mon-Fri, 0700-1900
Legislation: Yes
Mandatory Membership: No
Notice: 72 hrs. notice, no more than 10 days
Handle Emergency Notice? Yes
E-mail: jglyn.smith@sc1pups.org
E-mail 2: patsy.sharpe@sc1pups.org
Web Site: www.sc1pups.org

SOUTH DAKOTA

1-800-781-7474

SOUTH DAKOTA ONE CALL
Center # 1-800-873-3588
Alternate Center # 412-415-5000
115 Evergreen Heights Drive
Pittsburgh, PA 15229
Contact: Larry Englerth,
Executive Director
Contact # 605-339-0529
Fax # 605-339-0529
Do you accept FAX notifications?
Yes; 800-873-3954
Type Center: Contract
Coverage Area: Statewide
Members: 400
Hours of Operation: 24 hours
Legislation: Yes
Mandatory Membership: Yes
Notice: 48 hours, excluding weekends and holidays
Handle Emergency Notice? Yes
E-mail: sdonecall@qwest.net
Web Site: sdonecall.com

TENNESSEE

1-800-351-1111

TENNESSEE ONE-CALL SYSTEM, INC.
Center # 615-367-1110
Alternate Center # - 615-366-1987
2525 Perimeter Place Drive, Suite 118
Nashville, TN 37214
Contact: William B. Turner,
Executive Director
Contact # 615-367-1110
Fax # 615-366-5021
Do you accept FAX notifications?
Yes; 615-366-5021
Type Center: Self-Run
Coverage Area: Statewide
Members: 490
Hours of Operation: 24 hours
Legislation: Yes
Mandatory Membership: Mandatory for gas; Other utilities cannot collect damages if not a member.
Notice: No less than 3 working days nor more than 10 working days
Handle Emergency Notice? Yes
E-mail: bturner@tnonecall.com
Web Site: www.tnonecall.com

TEXAS

TEXAS ONE CALL SYSTEM
Center # 1-800-245-4545
(Nationwide)
Alternate Center # 412-415-5000
115 Evergreen Heights Drive
Pittsburgh, Pennsylvania 15229
Contact: Gary Craig, Director of Operations
Contact # 281-970-0505
Fax # 281-970-0110
Do you accept FAX notifications?
Yes; 800-217-3720
Type Center: Contract
Coverage Area: Statewide
Members: 205
Hours of Operation: 24 hours
Legislation: Yes
Mandatory Membership: Yes, ex-

cludes water, sewer and slurry
Notice: 2 working days, not more than 10 working days
Handle Emergency Notice? Yes
E-mail: gcraig@1-call.com
E-mail 2: txticket@1-call.com
Web Site: texasonecall.com

TEXAS EXCAVATION SAFETY SYSTEM, INC. (TESS)
Center # 1-800-344-8377
Alternate Center #
11880 Greenville Avenue, Ste. 120
Dallas, TX 75243-3568
Contact: Lee P. Marrs, President
Contact # 972-231-4597
Fax # 972-231-7006
Do you accept FAX notifications?
Yes; 1-800-690-1291
Type Center: Self-run
Coverage Area: Statewide
Members: 598
Hours of Operation: 24 hours
Legislation: Yes
Mandatory Membership:
For "Class A" facilities
Notice: 2 working days
Handle Emergency Notice? Yes
E-mail: leemarrs@digtess.org
Web Site: www.digtess.org

LONE STAR NOTIFICATION CENTER
Center # 1-800-669-8344
Alternate Center # 713-223-4567 (Local)
6575 West Loop South - Suite 225
Bellaire, TX 77401
Contact: Matt Ruddo,
Director of Marketing
Contact # 713-432-0365
Fax # 713-432-0041
Do you accept FAX notifications?
Yes; 713-432-0998
Type Center: Contract
Coverage Area: Statewide
Members: 155
Hours of Operation: 24 hrs./7 days
Legislation: Yes
Mandatory Membership: Yes
Notice: 2 working days
Handle Emergency Notice? Yes
E-mail: mcruddo@earthlink.net

UTAH
1-800-662-4111

BLUE STAKES OF UTAH UTILITY NOTIFICATION CENTER, INC.
Center # 1-800-662-4111
Alternate Center #
205 West 700 South - Suite 101
Salt Lake City, UT 84101
Contact: W. Gary Hansen,
Executive Director
Contact # 801-539-8683
Fax # 801-530-0562
Do you accept FAX notifications?
Yes; 801-530-0562
Type Center: Not-for-Profit
Coverage Area: Statewide
Members: 296
Hours of Operation: 0700-1700 M-F
Legislation: Yes
Mandatory Membership: Yes
Notice: 2 working days, 48 hrs. notice
Handle Emergency Notice? Yes
E-mail: garyh@bluestakes.org
Web Site: www.bluestakes.org

VERMONT
1-888-DIG-SAFE

DIG SAFE SYSTEM, INC.
Center # 781-721-0990
Alternate Center # 1-877-DIG-SAFE (1-877-344-7233)
331 Montvale Ave.
Woburn, MA 01801
Contact: Robert S. Finelli,
Executive Director
Contact # 781-721-1191
Fax # 781-721-0047
Do you accept FAX notifications?
Yes; 781-721-0047
Type Center: Not-for-Profit
Coverage Area: Statewide
Members: 380
Hours of Operation: 0600-1800 M-F
Legislation: Yes
Mandatory Membership: Yes
Notice: 48 hours, excluding weekends and holidays
Handle Emergency Notice? Yes
Web Site: www.digsafe.com

VIRGINIA

MISS UTILITY OF VIRGINIA
Center # 1-800-552-7001
Alternate Center # 804-530-1600
204 Rivers Bend Blvd
Chester, VA 23836
Contact: Philip Thompson,
General Manager
Contact # 804-530-1600
Fax # 804-530-2179
Do you accept FAX notifications?
Yes; 804-530-2179
Type Center: Contract
Coverage Area: Statewide (except Northern Virginia and Eastern Shore)
Members: 257
Hours of Operation: 0700-1700
Legislation: Yes
Mandatory Membership: Yes
Notice: 2 working days (48 hr notice required), Good for 15 working days
Handle Emergency Notice? Yes
E-mail: miss-utility@onisi.com
Web Site: www.onisi.com/onisi/msutil/default.htm

MISS UTILITY (Northern Virginia)
Center # 1-800-257-7777
Alternate Center #
7223 Parkway Drive
Hanover, MD 21076
Contact: Jim Holzer,
Director of Operations
Contact # 301-621-1317
Fax # 301-621-1711
Do you accept FAX notifications? Yes
301-621-1711
Type Center: Contract
Coverage Area: Northern Virginia
Members: 200
Hours of Operation: 0700-1700
Legislation: Yes
Mandatory Membership: Yes
Notice: 48 hours
Handle Emergency Notice? Yes
Web Site: www.missutility.net

WASHINGTON
1-800-424-5555

CALL BEFORE YOU DIG
Center # 1-800-424-5555
Call 1-877-668-4001 for addresses.
Contact # 877-668-4001
Do you accept FAX notifications? Yes
Type of Center: Contract
Coverage Area: Statewide
Members: 1,265
Hours of Operation: 24 hrs./365 days
Legislation: Yes
Mandatory Membership: Yes
Notice: 2 business days
Handle Emergency Notice? Yes
E-mail & Web Site: Call 1-877-668-4001

WEST VIRGINIA
1-800-245-4848

MISS UTILITY OF WEST VIRGINIA, INC.
Center # 1-800-245-4848
Alternate Center # 412-415-5065
2323 Windham Road
S. Charleston, WV 25303
Contact: Jim Kovach, Liaison Manager
Contact # 304-345-3959
Fax # 304-345-1851
Do you accept FAX notifications?
Yes; 800-217-3720
Type Center: Contract
Coverage Area: Statewide
Members: 274
Hours of Operation: 24 hours
Legislation: Yes
Mandatory Membership: Yes (with some exceptions)
Notice: 2 days; not more than 10 days
Handle Emergency Notice? Yes
Web Site: www.muwv.org

WISCONSIN
1-800-242-8511

DIGGERS HOTLINE, INC.
Center # 1-800-242-8511
Alternate Center # 414-259-0676
8112 West Bluemound Road
Wauwatosa, WI 53213-3356
Contact: Jeffrey D. Martens, Vice President Operations
Contact # 414-259-0676, ext. 203
Fax # 414-259-1453
Do you accept FAX notifications?
Yes; 1-800-338-3860
Type Center: Contract
Coverage Area: Statewide
Members: 965
Hours of Operation: 24 hours/7 days
Legislation: Yes
Mandatory Membership: Yes
Notice: 3 working days
Handle Emergency Notice? Yes
E-mail: jmartens@diggershotline.com
Web Site: www.diggershotline.com

WYOMING

ONE-CALL OF WYOMING
Center # 1-800-359-0249
AND 800-348-1030
You MUST call BOTH numbers per Wyoming law
PO Box 643
Casper, WY 82602-0643
Contact: Russell Waldner, President
Contact # 307-326-5260
Fax # 307-326-5934
Do you accept FAX notifications?
Yes, 307-472-3407 & 800-316-3753
You MUST fax BOTH numbers
Type Center: Non Profit; Member Utility
Coverage Area: Entire State
Members: 350
Hours of Operation: 24 hours
Legislation: Yes
Mandatory Membership: Yes
Notice: 2 full business days
Handle Emergency Notice? Yes
E-mail: rwaldner@carbonpower.com

DISTRICT OF COLUMBIA
1-800-257-7777

MISS UTILITY
Center # 1-800-257-7777
Alternate Center #
7223 Parkway Drive
Hanover, MD 21076
Contact: Jim Holzer, Director of Operations
Contact # 301-621-1317
Fax # 301-621-1711
Do you accept FAX notifications?
Yes; 301-621-1711
Type Center: Contract
Coverage Area: District of Columbia
Members:
Hours of Operation: 0700-1700
Legislation: Yes
Mandatory Membership: Yes
Notice: 2 days
Handle Emergency Notice? Yes

CANADA

ALBERTA ONE-CALL CORPORATION
Center # 1-800-242-3447
Alternate Center # 403-531-3700 (administration) 230,
6940 Fisher Road SE
Calgary, Alberta T2H 0W3 CANADA
Contact: Robert R. Chisholm, President
Contact # 403-531-3712
Fax # 403-531-3703
Do you accept FAX notifications?
Yes; 800-940-3447
Type Center: Non Profit, Staff
Coverage Area: Province of Alberta
Members: 384
Hours of Operation: 0600-2000/7 days
Legislation: No
Mandatory Membership: No
Notice: 2 full working days
Handle Emergency Notice?
Yes (24 hours/7 days)
E-mail: bchisholm@alberta1call.com
Web Site: alberta1call.com

ONTARIO ONE-CALL SYSTEM
Center # 1-800-400-2255 OR 905-709-1717
Alternate Center # 905-709-2891 (Administration line)
30 E. Beaver Creek Road, Ste. 103
Richmond Hill, Ontario L4B 1J2
Contact: Bill Roberts, Administrator
Contact # 888-251-9521
Fax # 888-251-9625
D o you accept FAX notifications?
Yes; 800-400-8876 OR 905-709-1711
Type Center: Contract
Coverage Area: Province of Ontario
Members: 75
Hours of Operation: 24 hours
Legislation: No
Mandatory Membership: No
Notice: Minimum 2 full business

days for standard requests
Handle Emergency Notice? Yes
E-mail: (under construction)
Web Site: www.on1call.com

INFO-EXCAVATION
Center # 1-800-663-9228
OR 514-286-9228
Alternate Center #
1600 Henri-Bourassa W., Suite 340
Montreal QC Canada H3M 3E2
Contact: Robert Jutras,
General Manager
Contact # 514-331-0635
Fax # 514-331-0791
Do you accept FAX notifications?
Yes; 1-800-441-3323 (plans & sketches accepted)
Type Center: Nonprofit
Coverage Area: Province of Quebec
Members: 47; Associates: 38
Hours of Operation: 24 hrs.
Legislation: No
Mandatory Membership: No
Notice: 72 hours (3 working days)
Handle Emergency Notice? Yes
E-mail: rjutras@info-ex.com
Web Site: www.info-ex.com
Other: Digital mapping service; Operations are ISO 9002 certified

BC ONE CALL
Center # 1-800-474-6886
Alternate Center # 1-604-257-1940
#222, 4259 Canada Way
Burnaby, BC V5G 1H1
Contact: Scott P. Henley
Contact # 604-257-1900
Fax # 604-451-0344
Do you accept FAX notifications?
Yes; 604-451-0344
Type Center: Contract
Coverage Area: British Columbia
Members: 55
Hours of Operation: 0700-1700
Legislation: No
Mandatory Membership: No
Notice: 2 working days
Handle Emergency Notice? Yes
E-mail: info@bconecall.bc.ca
Web Site: www.bconecall.bc.ca

AUSTRALIA

1100 (Toll free, nationwide)
Editor's Note: All One-Call services in Australia are part of the Association of Australian Dial Before You Dig Services (AADBYDS). For international calls, precede toll free number with 61. The national fax number is 1300 652 077. Precede number with 61-8 for international faxes. The AADBYDS internet web site is www.dialbeforeyoudig.com.au.

MELBOURNE ONE CALL SERVICE, INC.
Center # 1100 or 1300 652 044
Administration office # 61 3 9217 2833
P.O. Box 1276 Bundoora LPO
Bundoora VIC 3083 Australia
Contact: Mark Scopece,
IT Manager
(E-mail: mscopece@mocs.com.au)
Joe Maricchiolo, Management
(E-mail: jmaricch@mocs.au)
Contact # 61 3 9217 2819 (Mark);
61 3 9217 2818 (Joe)
Fax # 61 3 9217 2810
Do you accept FAX notifications?
Yes 61 3 1300 652 077
Type Center: Member Utility/Contractor
Coverage Area: States of Victoria, Tasmania, New South Wales, South Australia; Territories of Australian Capitaland Northern
Members: 90
Hours of Operation: 0800 - 1700 M-F
Legislation: No
Mandatory Membership: No
Notice: 2 working days
Handle Emergency Notice? Yes
E-mail: mocsinfo@mocs.com.au

DIAL BEFORE YOU DIG WA LTD.
Center # 1100 (Free Call)
Alternate Center #
C/- Water Corporation
13 South Street
Canning Vale, Western Australia 6155
Contact: Kim Rushton, Manager
Contact # 61-8-9424 8116
Fax # 61-8-9424 8122
Do you accept FAX notifications? Yes
Type Center: Nonprofit Staff
Coverage Area: Western Australia
Members: 31 (paying); 33 (non-paying)
Hours of Operation: 0800-1700
Legislation: No
Mandatory Membership: No
Notice:
Handle Emergency Notice? No
E-mail:
kim.rushton@watercorporation.com.au

SYDNEY ONE CALL SERVICE INC.
Trading as Dial Before You Dig NSW
Center # 1100 (within Australia)
Alternate Center # (03) 9217-2946 (outside Australia)
Center FAX # 1300-652-077 (within Australia)
Alternate Center FAX # (03) 9217-2719 (outside Australia)
535 Macdonalds Road, South Morang Victoria 3752 Australia
Contact: Dominic Puiu,
General Manager
Contact # 61-2-9365-7582
Fax # 61-2-9300-6482
Do you accept FAX notifications? Yes
Type Center: Contractor,
Melbourne One-Call Service Inc.
Coverage Area: New South Wales and Australian Capital Territority
Members: 16
Hours of Operation: 0800-1700 M-F
Legislation: None
Mandatory Membership: No
Notice: 2 working days normally required
Handle Emergency Notice? Yes
E-mail: mocsinfo@mocs.com.au (center); dpuiu@bigpond.com (contact)

DIAL BEFORE YOU DIG QUEENSLAND
Center # 13 21 29
(within Qld.; for location enquiries only)
Alternate Center # None
17 Lissner Street
Toowong, Q 4066
Mail address -P. O. Box 2022
Milton BC Australia 4064
Contact: Brian Zeller,
General Manager
Contact # 61 7 3511 5558
Fax # 61 7 3360 7786
Do you accept FAX notifications? Yes

(also E-mail from web site)
Type Center: Non profit contract for referral only (clearing house).
Coverage Area: Whole of Queensland
Members: 27
Hours of Operation: 0800-1700 M-F (except public holidays)
Legislation: No
Mandatory Membership: No
Notice: 2 working days for routine requests. Same day for urgent requests. 30 minutes for emergencies
Handle Emergency Notice? Yes
E-mail: bzqcbyd@ozemail.com.au

DIAL BEFORE YOU DIG SA/NT INC.
Center # 1100 (Australia Wide)
Alternate Center # n/a
Center FAX #
Address: P.O. Box 29 Lonsdale, SA 5160, Australia
Contact: Grant Mattiske (Chairman)
Phone # +61 8 8353 9113
Fax # +61 8 8235 0183
Mobile # +61 0419 825 783
Contact: Tom Faint (Manager SA/NT)
Phone # +61 8 8186 3301
Fax # +61 8 8326 4435
Mobile # +61 0414 265 597
Do you accept FAX notifications? Yes, +61 1300 652 077
Type Center: Referral center. Takes information from requestor and passes on to members for action. Members directly respond to requestor.
Coverage Area: All of South Australia and Northern Territory
Members: 40, plus 9 Associate Members and 14 Trial Members
Hours of Operation: 0800 – 1700 Mon-Fri
Legislation: None
Mandatory Membership: No
Notice: 2 clear working days
Handle Emergency Notice? Yes, response within 1 hour
E-mail 1: grant.mattiske@ndcglobal.com
E-mail 2: tfaint@ozemail.com.au
Web Site: www.dialbeforeyoudig.com.au

FINLAND

011-358-02-271-1181

JOHTOTIETO OY
Center # 011-358-09-271-1181
Alternate Center # 011-358-014-610288
Seutulantie 12
04401 Järvenpää, Finland
Contact: Pentti Halinen
Contact # 011-358-9-2711826
Fax # 011-358-9-2711977
Do you accept FAX notifications? Yes
Faksi: 09-27971120
Website: www.johtotieto.fi
Type Center: Non Profit & Contract
Coverage Area: Southern and Central parts of Finland
Members: 45
Hours of Operation: 0800 - 12600
Legislation: No
Mandatory Membership: No
Notice: 3 days
Handle Emergency Notice? Yes (only during office hours)

REPUBLIC OF CHINA

02 -351-2345

DIG CENTER, DIRECTORATE GENERAL OF TELECOMMUNICATIONS
Center # 02 -351-2345
Alternate Center #
31 Ai-Kuo East Road - Taiwan
Republic of China, 10605
Contact: Linda Lin, Vice Administrator
Do you accept FAX notifications? No
Type Center:
Coverage Area:
Members:
Hours of Operation: 0830-1730
Legislation: Yes
Mandatory Membership: No
Notice:
Handle Emergency Notice? No

SCOTLAND

800-800-333

SUSIEPHONE
Center # 800-800-333 (Scotland Only)
Alternate Center #
British Telecoms - 97 Caroline Terrace
Edinburgh, Scotland EH1 28QX
Contact:
Contact # 011-44-1-31-220-1234
Do you accept FAX notifications? No
Type Center: Contract
Coverage Area:
Members:
Hours of Operation: 0800-1700
Legislation: Yes
Mandatory Membership: No
Notice:
Handle Emergency Notice? No

UNITED KINGDOM

www.moleseye.com
MOLESEYE ONE CALL
This online roadworks register, based on and working on behalf of Scotland's Susiephone, serves utility companies, highway authorities, private contractors and individuals (anyone involved in construction planning, excavating, etc.) throughout the United Kingdom.

CURRENT DAMAGE PREVENTION LAWS (as known 8/6/02)

For more comprehensive, online information, visit the APWA website at: www.apwa.net/about/pet/1-call/

Be sure to scroll to the bottom of the page to "State Statutes."

ALABAMA

LEGISLATION: Yes, 94-487
DATE: 1994, enacted January 1995 PARTICIPATION: Volunteer
PROTECTS: All Utilities
POSITIVE RESPONSE: No, except in the case of where underground facilities are found that have not been marked and excavator initiates a callback. Utilities have up to 4 hours to positively respond.
EXEMPTS: D.O.T., County Officials, Railroads, Government Entity with Right-of-Way, Agricultural activities less than 12 inches in depth, Operating Water and Sewer Boards, Rural Water System and Property Owners or Easements.
ACCURACY OF LOCATION: 18" on either side
EMERGENCY CLAUSE: Yes
NOTIFICATION TIME: 2-10 working days, no more than 30 days prior to excavation COLOR CODE: Yes
PROTECTION/EXCAVATION: Yes
PENALTY CLAUSE: Yes
ISSUE PERMITS: No

ALASKA

LEGISLATION: AS 42.30
DATE: 09/01/98
PARTICIPATION: Volunteer
PROTECTS: All participating utilities
POSITIVE RESPONSE: No
EXEMPTS: No
ACCURACY OF LOCATION: 2' on either side
EMERGENCY CLAUSE: Yes
NOTIFICATION TIME: 2 working days minimum
COLOR CODE: Yes PROTECTION/EXCAVATION: Reasonable Care
PENALTY CLAUSE: $50-$1,000/offense
ISSUE PERMITS: No

ARIZONA

LEGISLATION: Article 6.3 Chapter 2 Section 40-360.21-.32
DATE: 1974, rev. 1989, 1996,1997, 1999
PARTICIPATION: Mandatory
PROTECTS: All Utilities
POSITIVE RESPONSE: Yes
EXEMPTS: Agricultural/Extraction of Natural Resources; Property Owners w/ hand tools while gardening or tilling; Emergency
ACCURACY OF LOCATION: 24" either side
EMERGENCY CLAUSE: Yes NOTIFICATION TIME: 2 working days, excluding weekends and state recognized holidays
COLOR CODE: Yes
PROTECTION/EXCAVATION: Hand dig, manually expose, support and protect within 24" either side of facility marking
PENALTY CLAUSE: $5,000 each violation
ISSUE PERMITS: No

ARKANSAS

LEGISLATION: Yes, Act 600 as amended in 1989, 1991, and 1995
DATE: 1987
PARTICIPATION: Mandatory (exceptions for some water, wastewater & CATV) PROTECTS: All Utilities
POSITIVE RESPONSE: Yes
EXEMPTS: (1) Agricultural - when not in Right-of-Way; (2) pre-engineered projects, (3) routine maintenance by state and counties, (4) cemeteries, (5) not using mechanized equipment or explosives.
ACCURACY OF LOCATION: 18" either side
EMERGENCY CLAUSE: Yes NOTIFICATION TIME: 2-10 working days
COLOR CODE: Yes
PROTECTION/EXCAVATION: Uncover using methods approved by facility owner
PENALTY CLAUSE: $2,500 each violation, $25,000 if gas or hazardous liquids involved.
ISSUE PERMITS: No

CALIFORNIA

LEGISLATION: Yes, GC 4216
DATE: 1984
PARTICIPATION: Mandatory PROTECTS: All underground facilities
POSITIVE RESPONSE: Yes
EXEMPTS: D. O. T.; Non-pressurized sewer, storm or drain lines
ACCURACY OF LOCATION: 24" either side
EMERGENCY CLAUSE: Yes
NOTIFICATION TIME: 2 working days but not more than 14 calendar days
COLOR CODE: Yes
PROTECTION/EXCAVATION: Expose with hand tools
PENALTY CLAUSE: $10,000 - $50,000; Loss of License by Contractor Board
ISSUE PERMITS: Yes, must have valid inquiry identification number from center to validate permits

COLORADO

LEGISLATION: Yes, S.B. 93-155
DATE: 1993
PARTICIPATION: Mandatory
PROTECTS: All Utilities
POSITIVE RESPONSE: YES
EXEMPTS: Railroad, D.O.T.
ACCURACY OF LOCATION: 18" on both sides of marking
EMERGENCY CLAUSE: Yes
NOTIFICATION TIME: 2 days not to include the day of actual notice
COLOR CODE: Yes
PROTECTION/EXCAVATION:
PENALTY CLAUSE: Yes
ISSUE PERMITS: No

CONNECTICUT

LEGISLATION: Yes, Public Act 77-16-345-9 G.C.R.
DATE: 1977, 1987
PARTICIPATION: Mandatory
PROTECTS: All Utilities
POSITIVE RESPONSE: Yes
EXEMPTS: None
ACCURACY OF LOCATION: 18" either side
EMERGENCY CLAUSE: Yes
NOTIFICATION TIME: 2 days
COLOR CODE: Yes
PROTECTION/EXCAVATION: If gas, hand dig within 18"
PENALTY CLAUSE: Up to $10,000
ISSUE PERMITS: No

DELAWARE

LEGISLATION: Yes, Title 26, Chapter 8
DATE: 1979
PARTICIPATION: Mandatory
PROTECTS: All Utilities and the Public Health and Safety
POSITIVE RESPONSE: Yes
EXEMPTS: Homeowners; Agricultural
ACCURACY OF LOCATION: Within 18"
EMERGENCY CLAUSE: Yes
NOTIFICATION TIME: 2-10 days
COLOR CODE: Yes
PROTECTION/EXCAVATION: Careful and prudent
PENALTY CLAUSE: $100 to $1000
ISSUE PERMITS: No
DIST. OF COLUMBIA
LEGISLATION: Yes, 3-240
DATE: 1979
PARTICIPATION: Mandatory
PROTECTS: All Utilities
POSITIVE RESPONSE: Yes
EXEMPTS: U.S. & D.C Governments
ACCURACY OF LOCATION: 18"
EMERGENCY CLAUSE: Yes
NOTIFICATION TIME: 2-10 days
COLOR CODE: Yes
PROTECTION/EXCAVATION: If gas, hand dig in

proximity unless D.C. has previously dug test hole
PENALTY CLAUSE: Damage without notification, double damages can be recovered $1000
ISSUE PERMITS: Yes

FLORIDA

LEGISLATION: Yes, Chapter 556.101-111
DATE: 1993
PARTICIPATION: Mandatory, except small cities
PROTECTS: All underground facilities
POSITIVE RESPONSE: Yes
EXEMPTS: Various, Refer to Chapter 556 Section 108 and 109
ACCURACY OF LOCATION: 24" either side
EMERGENCY CLAUSE: Yes NOTIFICATION TIME: Not less than 2 nor more than 5 business days
COLOR CODE: Yes
PROTECTION/EXCAVATION: Excavate in careful and prudent manner PENALTY CLAUSE: Non-criminal infraction (minimum fine $250-$5,000) ISSUE PERMITS: No

GEORGIA

LEGISLATION: Yes
DATE: 2000
PARTICIPATION: Mandatory
PROTECTS: All Utilities
POSITIVE RESPONSE: Yes
EXEMPTS: D.O.T. when replacing an existing structure that does not go below original grade
ACCURACY OF LOCATION: 24" from the outside edge
EMERGENCY CLAUSE: Yes
NOTIFICATION TIME: 48 hours excluding weekends and holidays
COLOR CODE: Yes
PROTECTION/EXCAVATION:
PENALTY CLAUSE: Up to $10,000 with Public Service Commission as enforcement agent
ISSUE PERMITS: No

HAWAII

LEGISLATION: No
DATE: In Committee
PARTICIPATION:
PROTECTS: All underground facilities
POSITIVE RESPONSE: Yes
EXEMPTS: None
ACCURACY OF LOCATION: 24" either side
EMERGENCY CLAUSE: Yes
NOTIFICATION TIME: 2 working days but not more than 14 calendar days
COLOR CODE: Yes
PROTECTION/EXCAVATION: Expose with hand tools
PENALTY CLAUSE: No
ISSUE PERMITS: No
NOTE: There is no legislation in Hawaii; however, the information above is used for operational procedures for the One-Call Center.

IDAHO

LEGISLATION: Yes, Chapter 22, Section 55-2201 through 55-2210
DATE: 2002 Revised
PARTICIPATION: Mandatory
PROTECTS: All Underground Utilities POSITIVE RESPONSE: No
EXEMPTS: Excavation of less than 15"; Agricultural; Extraction of Minerals; Normal road maintenance; Public Highway r/w; Railroads, except road crossings
ACCURACY OF LOCATION: within 24"
EMERGENCY CLAUSE: Yes
NOTIFICATION TIME: No more than 10 business days and no less than 2 business days
COLOR CODE: Yes
PROTECTION/EXCAVATION: Hand dig 24" either side of buried facility
PENALTY CLAUSE: Liable damages - Civil Penalty to $10,000
ISSUE PERMITS: No

ILLINOIS

LEGISLATION: Yes DATE: 1991, new amendatory language effective July 2002
PARTICIPATION: Mandatory PROTECTS: All Utilities
POSITIVE RESPONSE: Yes
EXEMPTS: Agricultural; Railroads
ACCURACY OF LOCATION: 18" EMERGENCY CLAUSE: Yes
NOTIFICATION TIME: Illinois state law requires at least 48-hour notice (two working days), but no more than a 14-calendar day advance notice before start of excavation
COLOR CODE: Yes
PROTECTION/EXCAVATION:
PENALTY CLAUSE: Effective July 2002, up to $5,000 when no call is made; up to $2,500 when call is made, but excavator does not comply with other portions of Act; up to $5,000 for each separate utility offense. Further, adds language to establish penalties up to $2,500 for excavators who call in emergency requests that are not defined emergencies in the Act; up to $1,000 for individuals not involved in process who remove markings before the completion of a project. Administrative penalty program through the Illinois Commerce Commission.
ISSUE PERMITS: No

INDIANA

LEGISLATION: Yes
DATE: 1990, amended 1994
PARTICIPATION: Voluntary
PROTECTS:
POSITIVE RESPONSE: Yes
EXEMPTS: Normal farming practices, hand digging, RR except at public grade crossings
ACCURACY OF LOCATION: 24" on either side of the buried facility
EMERGENCY CLAUSE: Yes
NOTIFICATION TIME: 2 full working days, but no more than 20 calendar days
COLOR CODE: Yes
PROTECTION/EXCAVATION: Hand dig 24" on either side of the buried facility
PENALTY CLAUSE: Damages, plus up to three times punitive damages
ISSUE PERMITS: No

IOWA

LEGISLATION: Yes
DATE: 1/1/93
PARTICIPATION: Mandatory PROTECTS: All underground facilities
POSITIVE RESPONSE: Yes
EXEMPTS: None
ACCURACY OF LOCATION: 18" either side
EMERGENCY CLAUSE: Yes NOTIFICATION TIME: 2 working days
COLOR CODE: Yes
PROTECTION/EXCAVATION:
PENALTY CLAUSE: Yes
ISSUE PERMITS: No

KANSAS

LEGISLATION: Yes, Statute 66-1801-66-1814 - 07-01-93
DATE: 1993 PARTICIPATION: Mandatory
PROTECTS: All utilities (Except water and sewer)
POSITIVE RESPONSE: Yes
EXEMPTS: Water and sewer
ACCURACY OF LOCATION: Within 24"
EMERGENCY CLAUSE: Yes NOTIFICATION TIME: 2-10 working days
COLOR CODE: Yes
PROTECTION/EXCAVATION: Yes
PENALTY CLAUSE: Yes
ISSUE PERMITS: No

KENTUCKY

LEGISLATION: Yes, HB 613 effective 1/1/95
DATE: 1994
PARTICIPATION: Volunteer
PROTECTS: All utilities
POSITIVE RESPONSE: Yes
EXEMPTS: Excavation by operator on own easement; Routine road maintenance; Non-powered hand tools; Animal power; Agricultural; Residential Property; Grave in cemetery; Solid waste disposal site; Coal mining
ACCURACY OF LOCATION: 18" either side
EMERGENCY CLAUSE: Yes
NOTIFICATION TIME: 2 business days
COLOR CODE: Yes
PROTECTION/EXCAVATION: Careful and prudent manner
PENALTY CLAUSE: Fines; cost of repair
ISSUE PERMITS: No

LOUISIANA

LEGISLATION: Yes, ACT 923
DATE: 1988, Amended 1992, 1995, 1997, 1999, 2001
PARTICIPATION: Mandatory must participate in a regional notification program POSITIVE RESPONSE: Utility operators must mark or provide information regarding location of underground or submerged utilities.
PROTECTS: All owners/operators of UG and submerged facilities who participate in damage prevention program
EXEMPTS: Cable TV from participation, but must call in locate requests. Political subdivisions if they have opted out, but must call in locate requests
ACCURACY OF LOCATION: 18" either side
EMERGENCY CLAUSE: Yes
NOTIFICATION TIME: 48 to 120 hours in advance
COLOR CODE: Yes
PROTECTION/EXCAVATION: Yes
PENALTY CLAUSE: Yes
ISSUE PERMITS: No

MAINE

LEGISLATION: Yes, Chapter 437 DATE: 1991
PARTICIPATION: Mandatory
PROTECTS: Gas, Oil, Telephone, Electric, Cable TV
POSITIVE RESPONSE: No
EXEMPTS: None
ACCURACY OF LOCATION: 36" EMERGENCY CLAUSE: Yes
NOTIFICATION TIME: 72 hours; excluding weekends and holidays
COLOR CODE: Yes
PROTECTION/EXCAVATION:
PENALTY CLAUSE: up to $1000 ISSUE PERMITS: Yes
PRE-MARK REQUIREMENT: Yes

MARYLAND

LEGISLATION: Yes, Article 78, Section 28A
DATE: 1974
PARTICIPATION: Mandatory
PROTECTS: All Utilities
POSITIVE RESPONSE: Yes
EXEMPTS: None
ACCURACY OF LOCATION: 18" either side
EMERGENCY CLAUSE: Yes
NOTIFICATION TIME: 48 hours
COLOR CODE: Yes
PROTECTION/EXCAVATION: Excavate in careful and prudent manner
PENALTY CLAUSE: $1000 or 10 times the cost of damage
ISSUE PERMITS: No

MASSACHUSETTS

LEGISLATION: Yes, Chapter 82, Section 40
DATE: 1980
PARTICIPATION: Mandatory
PROTECTS: Gas, Electric, Telephone, CATV and Private Water Companies
POSITIVE RESPONSE: No
EXEMPTS: None
ACCURACY OF LOCATION: None EMERGENCY CLAUSE: Yes NOTIFICATION TIME: 72 hours, exclusive of weekends and holidays
COLOR CODE: Yes
PROTECTION/EXCAVATION:
PENALTY CLAUSE: First $500, Subsequent $1000 to $5000
ISSUE PERMITS: Yes
PRE-MARK REQUIREMENT: Yes

MICHIGAN

LEGISLATION: Yes, Public Act 53 as Amended
DATE: 1989
PARTICIPATION: Mandatory
PROTECTS: All utilities
POSITIVE RESPONSE: Yes - Voluntary
EXEMPTS: Non-powered hand tools; D.O.T.
ACCURACY OF LOCATION: 18" either side
EMERGENCY CLAUSE: Yes NOTIFICATION TIME: 3 full working days COLOR CODE: No PROTECTION/EXCAVATION: Hand dig test holes
PENALTY CLAUSE: Up to $5,000 fine and/or 1 year in jail, injunctive relief. May be enjoined from work in state
ISSUE PERMITS: No

MINNESOTA

LEGISLATION: Yes, Chapter 216D
DATE: 1987
PARTICIPATION: Mandatory
PROTECTS: All Utilities
POSITIVE RESPONSE: No
EXEMPTS: Extraction of minerals; opening a grave; road grading that does not involve the ditch; normal farming operations less than 18"; gardening less than 12"; planting of windbreaks, shelterbelts, or tree plantations less than 18".
ACCURACY OF LOCATION: Within 24" either side
EMERGENCY CLAUSE: Yes
NOTIFICATION TIME: 48 hours
COLOR CODE: Yes
PROTECTION/EXCAVATION: Excavation in careful and prudent manner
PENALTY CLAUSE: Up to $1,000 fine per violation per day of violation. An excavator who knowingly damages a facility and does not notify the facility operator as soon as possible or backfills is guilty of a misdemeanor.
ISSUE PERMITS: No

MISSISSIPPI

LEGISLATION: Mississippi Statute 77-13-1
DATE: 1985
PARTICIPATION: Volunteer
PROTECTS: Buried utilities
POSITIVE RESPONSE: Yes
EXEMPTS: None
ACCURACY OF LOCATION: 18"
EMERGENCY CLAUSE: Yes
NOTIFICATION TIME:
COLOR CODE: Yes
PROTECTION/EXCAVATION:
PENALTY CLAUSE: No
ISSUE PERMITS: No

MISSOURI

LEGISLATION: Yes
DATE: 1991, revised 2001
PARTICIPATION: Mandatory
PROTECTS: All Utilities
POSITIVE RESPONSE: Yes
EXEMPTS: Agricultural & Railroads
ACCURACY OF LOCATION: 2' plus width of facility
EMERGENCY CLAUSE: Yes
NOTIFICATION TIME: 2 working days
COLOR CODE: Yes
PROTECTION/EXCAVATION: Excavate in careful and prudent manner
PENALTY CLAUSE: Yes
ISSUE PERMITS: No

MONTANA

LEGISLATION: Yes, Chapter 180, 6-69-4-501
DATE: 1971, 1991, 1997 PARTICIPATION: Mandatory
PROTECTS: All Utilities
POSITIVE RESPONSE: Yes
EXEMPTS: Surface road grading and ditch or road maintenance that does not change the original road or ditch grade or flow line.
ACCURACY OF LOCATION: 18" either side of facility
EMERGENCY CLAUSE: Yes
NOTIFICATION TIME: 2 business days
COLOR CODE: Yes
PROTECTION/EXCAVATION: Hand dig 18" either side of facilities.
PENALTY CLAUSE: Liable damages - Civil penalty
ISSUE PERMITS: No

NEBRASKA

LEGISLATION: Yes, One-Call Notification System Act Section 76-2301 - 76-2330
DATE: 1995
PARTICIPATION: Mandatory
PROTECTS: All operators of underground facilities
POSITIVE RESPONSE: Yes
EXEMPTS:
ACCURACY OF LOCATION: 18" either side
EMERGENCY CLAUSE: Yes NOTIFICATION TIME: 2-10 business days
COLOR CODE: Yes
PROTECTION/EXCAVATION:
PENALTY CLAUSE: Yes; Gas and hazardous liquids $10,000 per day maximum of $500,000;

Others $500 per day maximum of $5,000
ISSUE PERMITS: No

NEVADA

LEGISLATION: Yes, NRS 455.080-NRS 455.180
DATE: 1987
PARTICIPATION: Mandatory
PROTECTS: All underground facilities POSITIVE RESPONSE: Yes
EXEMPTS: D.O.T.
ACCURACY OF LOCATION: 30" either side
EMERGENCY CLAUSE: Yes
NOTIFICATION TIME: 2 working days but not more than14 calendar days
COLOR CODE: Yes
PROTECTION/EXCAVATION: Expose with hand tools
PENALTY CLAUSE: $1,000-$100,000
ISSUE PERMITS: No

NEW HAMPSHIRE

LEGISLATION: Yes, RSA 374, Section 47 thru 56
DATE: 1983
PARTICIPATION: Mandatory
PROTECTS: Gas, Electric, Telephone, Cable TV and Public Water Utilities
POSITIVE RESPONSE: No
EXEMPTS: None
ACCURACY OF LOCATION: 18"
EMERGENCY CLAUSE: Yes NOTIFICATION TIME: 72 hours, exclusive of weekends and holidays
COLOR CODE: Yes
PROTECTION/EXCAVATION:
PENALTY CLAUSE: $500, Subsequent $5000
ISSUE PERMITS: Yes

NEW JERSEY

LEGISLATION: NJSA 48:2-74 thru 91 DATE: 1994
PARTICIPATION: Mandatory
PROTECTS: All Underground Facilities
POSITIVE RESPONSE: Yes
EXEMPTS: Non-Forced Sewers
ACCURACY OF LOCATION: 18" EMERGENCY CLAUSE: Yes
NOTIFICATION TIME: 3-10 days COLOR CODE: Yes
PROTECTION/EXCAVATION: Hand dig 2' either side of Underground Facility
PENALTY CLAUSE: Yes
ISSUE PERMITS: No

NEW MEXICO

LEGISLATION: Yes, NMSA Chapter 62, Article 14
DATE: 1997
PARTICIPATION: Pipelines (including gas distribution) in or near public right-of-way, Mandatory; Others - Voluntary
PROTECTS: All undergroud facilities
POSITIVE RESPONSE: No
EXEMPTS: None
ACCURACY OF LOCATION: 12" Horizontal
EMERGENCY CLAUSE: Yes NOTIFICATION TIME: 2 working days
COLOR CODE: Yes
PROTECTION/EXCAVATION: Maintain 18" from cutting edge
PENALTY CLAUSE: $500
ISSUE PERMITS: No

NEW YORK

LEGISLATION: Yes
DATE: February 2001
PARTICIPATION: Mandatory
PROTECTS: All undergroud facility owners/operators
POSITIVE RESPONSE: Yes
EXEMPTS: None
ACCURACY OF LOCATION: Within 2 feet of the edge of facility
EMERGENCY CLAUSE: Yes NOTIFICATION TIME: 2-10 business days
COLOR CODE: Yes
PROTECTION/EXCAVATION: Hand dig test holes; No power equipment within 4" of facility after verification.
PENALTY CLAUSE: Yes
ISSUE PERMITS: No

NORTH CAROLINA

LEGISLATION: Yes, Chapter 87-100 DATE: 1986
PARTICIPATION: Volunteer
PROTECTS: All Utilities
POSITIVE RESPONSE: No
EXEMPTS: None
ACCURACY OF LOCATION: 2.5' either side
EMERGENCY CLAUSE: Yes
NOTIFICATION TIME: 48 hours to 10 days
COLOR CODE: Yes
PROTECTION/EXCAVATION: Excavate in careful and prudent manner
PENALTY CLAUSE: No
ISSUE PERMITS: No

NORTH DAKOTA

LEGISLATION: ND Century Code Chapter 49.23
DATE: 1973, 1995
PARTICIPATION: Mandatory
PROTECTS: All utilities
POSITIVE RESPONSE: Yes
EXEMPTS: Opening a grave in a cemetery; Agricultural less than 18" in depth; Normal maintenance of roads and streets: Normal repair and maintenance of railroad right-of-way.
ACCURACY OF LOCATION: 2' on other side of the UG facility except for UG facilities
EMERGENCY CLAUSE: Yes, used to convey water.
NOTIFICATION TIME: 48 hours
COLOR CODE: Yes
PROTECTION/EXCAVATION: Avoid damage and minimize interference to underground facility. Careful and prudent manner.
PENALTY CLAUSE:
ISSUE PERMITS: Yes

OHIO

LEGISLATION: Yes, Ohio S.B. 174, 2 tier legislation 1 53.64, 3781.25-32
DATE: 1988, 1990
PARTICIPATION: Mandatory
PROTECTS: All Utilities
POSITIVE RESPONSE: No
EXEMPTS: Private Septic Systems; Mining; Agricultural less than 12" in depth; Property owners
ACCURACY OF LOCATION: 18"
EMERGENCY CLAUSE: Yes NOTIFICATION TIME: 48 hrs. to 10 days
COLOR CODE: Yes
PROTECTION/EXCAVATION:
PENALTY CLAUSE:
ISSUE PERMITS: No

OKLAHOMA

LEGISLATION: Yes, Title 63, Section 142-1-11, HB 1029-1982
DATE: 1982
PARTICIPATION: Voluntary except for gas and liquid pipelines
PROTECTS: All underground facilities
POSITIVE RESPONSE: Yes
EXEMPTS: Public agencies and their contractors in public right-of-way on pre-engineered project, certified project, or routine maintenance.
ACCURACY OF LOCATION: 2' on either side of underground facility
EMERGENCY CLAUSE: Yes
NOTIFICATION TIME: Not less than 48 hours or more than 10 days, excluding weekends and holidays
COLOR CODE: Yes
PROTECTION/EXCAVATION: Hand dig 2' either side of underground facility before using power equipment.
PENALTY CLAUSE: No
ISSUE PERMITS: By County and Municipal government

OREGON

LEGISLATION: Yes, ORS 952-001-0010 through 952-001-0090
DATE: 1987, 1995
PARTICIPATION: Mandatory
PROTECTS: All Utilities
POSITIVE RESPONSE: Yes
EXEMPTS: Private Property not within right of way or easement; Routine Road Maintenance less than 12"; Agricultural
ACCURACY OF LOCATION: 24" outside the edge of the facility
EMERGENCY CLAUSE: Yes
NOTIFICATION TIME: 2 days - life of the project. Marks maintained by Excavator.
COLOR CODE: Yes

PROTECTION/EXCAVATION: Use responsible care and hand dig
PENALTY CLAUSE: Yes
ISSUE PERMITS: No

PENNSYLVANIA

LEGISLATION: Yes, ACT 287 as amended by Act 187
DATE: 1974, 1996
PARTICIPATION: Mandatory
PROTECTS: All Underground Facility Owners
POSITIVE RESPONSE: Yes Response to all notices mandatory through One-Call Center
EXEMPTS: Agricultural less than 18"; D.O.T. less than 24" within state highway R-O-W and local municipalities less than 18" within road R-O-W; and extraction of natural resources
ACCURACY OF LOCATION: APWA Standards
EMERGENCY CLAUSE: Yes
NOTIFICATION TIME: Construction phase-3-10 working days; Design phase-10 to 90 working days
COLOR CODE: Yes
PROTECTION/EXCAVATION: Hand dig test holes; careful and prudent techniques within the tolerance zone
PENALTY CLAUSE: Min. $2,500, no more than $25,000 + up to 90 days in jail; citations for minor infractions ($100-$500 fines after third warning) enforced by the Department of Labor & Industry Bureau of Labor Law Compliance – 1-800-932-0665.
ISSUE PERMITS: Yes

RHODE ISLAND

LEGISLATION: Yes, Section 39- 1.2 DATE: 1984
PARTICIPATION: Mandatory
PROTECTS: Electric, Gas, Telephone and Cable TV and Public Water Companies
POSITIVE RESPONSE: No
EXEMPTS: None
ACCURACY OF LOCATION: 18"
EMERGENCY CLAUSE: Yes
NOTIFICATION TIME: 48 hours, exclusive of weekends and holidays
COLOR CODE: Yes
PROTECTION/EXCAVATION:
PENALTY CLAUSE: First $200, second $500, Subsequent $1,000
ISSUE PERMITS: Yes

SOUTH CAROLINA

LEGISLATION: Yes
DATE: 1978
PARTICIPATION: Volunteer PROTECTS: All utilities
POSITIVE RESPONSE: No
EXEMPTS: Conditional
ACCURACY OF LOCATION: 2.5' or 30" either side
EMERGENCY CLAUSE: Yes
NOTIFICATION TIME: 3 working days minimum; 10 days maximum.
COLOR CODE: Yes
PROTECTION/EXCAVATION: Plan excavation to avoid damage to or minimize interference; maintain clearance; provide support as required by operators.
PENALTY CLAUSE: $1,000 maximum ISSUE PERMITS: No

SOUTH DAKOTA

LEGISLATION: Yes, SDCL 49-7A
DATE: 1977, 1993, amended 1997 PARTICIPATION: Mandatory
PROTECTS: All Utilities
POSITIVE RESPONSE: No (Yes for Emergency Tickets Only
EXEMPTS: Tilling of soil and gardening up to 12"; Agricultural; Road and ditch maintenance; Grave in cemetery; Planned Sanitary Landfill
ACCURACY OF LOCATION: 18"
EMERGENCY CLAUSE: Yes
NOTIFICATION TIME: 48 hrs., excluding weekends and holidays
COLOR CODE: Yes
PROTECTION/EXCAVATION: Maintain 18" clearance from cutting edge
PENALTY CLAUSE:
ISSUE PERMITS: No

TENNESSEE

LEGISLATION: Yes, TCA 65-31-101 DATE: 1978, amended April 1999
PARTICIPATION: Gas utilities are mandated to join; other utilities cannot collect damages if they do not participate PROTECTS:
POSITIVE RESPONSE: Yes
EXEMPTS: None
ACCURACY OF LOCATION: 2' on either side of the facility plus the size of the facility EMERGENCY CLAUSE: Yes
NOTIFICATION TIME: 3-10 working days COLOR CODE: Yes
PROTECTION/EXCAVATION: Required to call
PENALTY CLAUSE: Up to $2,500 and/or 48 hours in jail
ISSUE PERMITS: No

TEXAS

LEGISLATION: Yes, Utilities Code Title 5, Article 251.
DATE: Oct. 1, 1998, Rev 9/1 & 11/1/2001
PARTICIPATION: Mandatory for all facilities except water, sewer and slurry
PROTECTS: "Class A" facilities
POSITIVE RESPONSE: Yes
EXEMPTS: DOT, water, sewer, slurry ACCURACY OF LOCATION:
EMERGENCY CLAUSE: Yes NOTIFICATION TIME: At least 2 working days, not more than 14 days
COLOR CODE: Yes PROTECTION/EXCAVATION:
PENALTY CLAUSE: Yes
ISSUE PERMITS: No

UTAH

LEGISLATION: Yes, U.C.A. 54-8a-1 - 54-8a-11
DATE: 1953, 1977, amended 2001
PARTICIPATION: Mandatory
PROTECTS: All utilities
POSITIVE RESPONSE: Yes
EXEMPTS: Emergencies, gardening and tilling private ground
ACCURACY OF LOCATION: 24"
EMERGENCY CLAUSE: Yes NOTIFICATION TIME: 14 days
COLOR CODE: Yes PROTECTION/EXCAVATION: Excavator shall proceed in a manner that is reasonably calculated to avoid damage to the underground facilities. Hand-dig in 24" tolerance zone until facilities are located.
PENALTY CLAUSE: Yes
ISSUE PERMITS: No

VERMONT

LEGISLATION: Yes, Chapter 890
DATE: 1980, 1994
PARTICIPATION: Mandatory
PROTECTS: All utilities
POSITIVE RESPONSE: Yes, excavator-operator ticket information exchange system EXEMPTS: Property owners hand digging; agricultural; railroads; routine road maintenance; mining; D.O.T.; emergency
ACCURACY OF LOCATION: Within 2' of either side
EMERGENCY CLAUSE: Yes NOTIFICATION TIME: 48 hours - good for 15 working days
COLOR CODE: Yes
PROTECTION/EXCAVATION: Shall take all reasonable steps necessary; hand digging
PENALTY CLAUSE: Yes
ISSUE PERMITS: No

VIRGINIA

LEGISLATION: Yes, Chapter 890
DATE: 1980, 1994
PARTICIPATION: Mandatory
PROTECTS: All Utilities
POSITIVE RESPONSE: Yes, Excavator-Operator Ticket Information Exchange System
EXEMPTS: Property Owners hand digging, Agricultural, Railroads, Routine road maintenance, Mining, D.O.T., Emergency
ACCURACY OF LOCATION: Within 2 feet either side
EMERGENCY CLAUSE: Yes
NOTIFICATION TIME: 48 hours - good for 15 working days
COLOR CODE: Yes
PROTECTION/EXCAVATION: Shall take all reasonable steps necessary; Hand digging
PENALTY CLAUSE: Yes
ISSUE PERMITS: No

WASHINGTON

LEGISLATION: Yes, RCW 19.122
DATE: 1984, 1990
PARTICIPATION: Mandatory
PROTECTS: All utilities
POSITIVE RESPONSE: No
EXEMPTS: Excavation of less than 12"
ACCURACY OF LOCATION: 2' on either side
EMERGENCY CLAUSE: Yes
NOTIFICATION TIME: 2 business days minimum, 10 business days maximum
COLOR CODE: Yes
PROTECTION/EXCAVATION: Excavation in a careful and prudent manner
PENALTY CLAUSE: $1,000 + treble damages, downtime; $10,000 for damaging hazardous liquid or gas pipeline without a call before digging
ISSUE PERMITS: No

WEST VIRGINIA

LEGISLATION: Chapter 24-C
DATE: July 15, 1996,
PARTICIPATION: Mandatory (with some exceptions)
PROTECTS: All participating MUWV members
POSITIVE RESPONSE: No
EXEMPTS: Private Water Companies, and State, Federal, County and City Entities
ACCURACY OF LOCATION: 2'
EMERGENCY CLAUSE: Yes
NOTIFICATION TIME: 2 business days, but not more than 10 days
COLOR CODE: Yes
PROTECTION/EXCAVATION: As stated by Chapter 24-C
PENALTY CLAUSE: No
ISSUE PERMITS: No

WISCONSIN

LEGISLATION: WI State Statute 182.0175
DATE: Jan. 1997
PARTICIPATION: Mandatory
PROTECTS: All Owners of Underground & Overhead Facilities
POSITIVE RESPONSE: No
EXEMPTS: None
ACCURACY OF LOCATION: 18"
EMERGENCY CLAUSE: Yes NOTIFICATION TIME: 3 working days
COLOR CODE: Yes
PROTECTION/EXCAVATION: Maintain 18" clearance from cutting edge
PENALTY CLAUSE: $2,000
ISSUE PERMITS: No

WYOMING

LEGISLATION: Yes, Wyoming Statute 37-12-301
DATE: 1996
PARTICIPATION: Mandatory PROTECTS: All Utilities
POSITIVE RESPONSE: Yes
EXEMPTS: Wyoming Statutes 37-12-305 A-G
ACCURACY OF LOCATION: 24" horizontally from exterior sides of facility
EMERGENCY CLAUSE: Yes NOTIFICATION TIME: 2-14 business days
COLOR CODE: Yes
PROTECTION/EXCAVATION: Excavation in careful and prudent manner
PENALTY CLAUSE: If damage occurs, up to $5,000 in Civil Penalty
ISSUE PERMITS: No

Future APWA Show Dates and Sites

INTERNATIONAL PUBLIC WORKS CONGRESS & EXPOSITION

YEAR	DATE	CITY	LOCATION
2003	August 24-27	San Diego, CA	Convention Center
2004	September 12-15	Atlanta,GA	GA World Congress Ctr.
2005	September 11-14	Minneapolis, MN	Minneapolis Conv. Ctr.
2006	September 10-13	New Orleans, LA	Morial Convention Ctr.
2007	September 9-12	San Antonio, TX	Henry B. Gonzalez Conv. Ctr.

OCSI EXECUTIVE COUNCIL AND DELEGATES COMMITTEE

OCSI members meet two times per year, in February and at APWA's International Public Works Congress and Exposition.

NATIONAL PUBLIC WORKS WEEK

2003	May 18-24
2004	May 16-22
2005	May 15-21
2006	May 14-20
2007	May 20-26

OCSI DIRECTORY INFORMATION AND CONTACT LIST

DIRECTORY OR LEGISLATIVE UPDATES AND/OR CHANGES CAN BE SENT AT ANY TIME, BUT ARE REQUIRED AT LEAST ANNUALLY. IF VERIFICATION IS NOT RECEIVED FOR TWO CONSECUTIVE YEARS, INFORMATION WILL BE DROPPED FROM THE DIRECTORY.

Send updates and/or changes to the APWA Publications Manager at the address listed below.

APWA, 2345 Grand Boulevard, Suite 500, Kansas City, MO 64108-2641
(816) 472-6100/ FAX (816) 472-1610
E-mail: bookstore@apwa.net

A downloadable version of this directory is available at www.apwa.net.

Glossary

ACCEPTED ENGINEERING PRACTICES: Procedures compatible with the standards of practice required of a registered professional engineer.

ADJACENT STRUCTURE STABILITY: The stability of the foundation(s) of adjacent structures whose location may create surcharges, changes in soil conditions, or other disruptions that have the potential to extend into the failure zone of the excavation or trench.

BELL-BOTTOM PIER HOLE: A type of shaft or footing excavation, the bottom of which is made larger than the cross section above to form a belled shape.

BENCHING SYSTEMS: A method of protecting employees from cave-ins by excavating the sides of an excavation to form one or a series of horizontal levels or steps, usually with vertical or near-vertical surfaces between levels.

BOILING: A condition evidenced by an upward water flow into the bottom of the cut. A high water table is one of the causes of boiling. Boiling produces a "quick" condition in the bottom of the cut, and can occur even when shoring or trench boxes are used.

CAVE-IN: The separation of a mass of soil or rock material from the side of an excavation, or the loss of soil from under a trench shield or support system, and its sudden movement into the excavation, either by falling or sliding, in sufficient quantity so that it could entrap, bury, or otherwise injure and immobilize a person.

COHESIVE SOIL: Clay (fine grained soil), or soil with a high clay content, that has cohesive strength. Cohesive soil does not crumble, can be excavated with vertical sideslopes, and is plastic when moist. Cohesive soil is hard to break up when dry, and exhibits significant cohesion when submerged. Cohesive soils include clayey silt, sandy clay, silty clay, clay, and organic clay.

COMPETENT PERSON: An individual who is capable of identifying existing and predictable hazards or working conditions that are hazardous, unsanitary, or dangerous to employees, and who has authorization to take prompt corrective measures to eliminate or control these hazards and conditions.

CONFINED SPACE: A space that, by design and/or configuration, has limited openings for entry and exit, unfavorable natural ventilation, may contain or produce hazardous substances, and is not intended for continuous employee occupancy.

CROSS BRACES: The horizontal members of a shoring system installed perpendicular to the sides of the excavation, the ends of which bear against either uprights or wales.

EXCAVATION: Any man-made cut, cavity, trench, or depression in an earth surface that is formed by earth removal. A trench is a narrow excavation (in relation to its length) made below the surface of the ground. In general, the depth of a trench is greater than its width, and the width (measured at the bottom) is not greater than 15 feet. If a form or other structure installed or constructed in an excavation reduces the distance between the form and the side of the excavation to 15 feet or less (measured at the bottom of the excavation), the excavation is also considered to be a trench.

FACES: The vertical or inclined earth surfaces formed as a result of excavation work.

FAILURE: The breakage, displacement, or permanent deformation of a structural member or connection so as to reduce its structural integrity and its supportive capabilities.

HAZARDOUS ATMOSPHERE: An atmosphere that by reason of being explosive, flammable, poisonous, corrosive, oxidizing, irritating, oxygen-deficient, toxic, or otherwise harmful may cause death, illness, or injury to persons exposed to it.

INGRESS AND EGRESS: Mean "entry" and "exit," respectively. In trenching and excavation operations, they refer to the provision of safe means for employees to enter or exit an excavation or trench.

PROTECTIVE SYSTEM: A method of protecting employees from cave-ins, from material that could fall or roll from an excavation face or into an excavation, and from the collapse of adjacent structures. Protective systems include support systems, sloping and benching systems, shield systems, and other systems that provide the necessary protection.

RAMP: An inclined walking or working surface that is used to gain access to one point from another, and is constructed from earth or from structural materials such as steel or wood.

REGISTERED PROFESSIONAL ENGINEER: A person who is registered as a professional engineer in the state where the work is to be performed. However, a professional engineer who is registered in any state is deemed to be a "registered professional engineer" within the meaning of Subpart P when approving designs for "manufactured protective systems" or "tabulated data" to be used in interstate commerce.

SHEETING: The members of a shoring system that retain the earth in position and in turn are supported by other members of the shoring system.

SHIELD (Shield system): A structure that is able to withstand the forces imposed on it by a cave-in and thereby protect employees within the structure. Shields can be permanent structures or can be designed to be portable and moved along as work progresses. Additionally, shields can be either pre-manufactured or job-built in accordance with 1926.652(c)(3) or (c)(4). Shields used in trenches are usually referred to as "trench boxes" or "trench shields."

SLOPING (Sloping system): A method of protecting employees by excavating to form sides of an excavation that are inclined away from the excavation so as to prevent cave-ins. The angle of incline required to prevent a cave-in varies with differences in such factors as the soil type, environmental conditions of exposure, and application of surcharge loads.

SHORING (Shoring system): A structure such as a metal hydraulic, mechanical, or timber shoring system that supports the sides of an excavation and is designed to prevent cave-ins.

Spall: A piece or section of concrete that has broken away from the rest of the concrete.

SPOIL: The dirt, rocks, and other materials removed from an excavation and either temporarily or permanently put aside.

STABLE ROCK: Natural solid mineral material that can be excavated with vertical sides and will remain intact while exposed. Unstable rock is considered to be stable when the rock material on the side or sides of the excavation is secured against caving-in or movement by rock bolts or by another protective system that has been designed by a registered professional engineer.

STRUCTURAL RAMP: A ramp built of steel or wood, usually used for vehicle access. Ramps made of soil or rock are not considered structural ramps.

SUPPORT SYSTEM: A structure such as underpinning, bracing, or shoring, that provides support to an adjacent structure, underground installation, or the sides of an excavation.

SUBSURFACE ENCUMBRANCES: Include underground utilities, foundations, streams, water tables, transformer vaults, and geological anomalies.

SURCHARGE: An excessive vertical load or weight caused by spoil, overburden, vehicles, equipment, or activities that may affect trench stability.

TABULATED DATA: Tables and charts approved by a registered professional engineer and used to design and construct a protective system.

TRENCH (Trench excavation): A narrow excavation (in relation to its length) made below the surface of the ground. In general, the depth is greater than the width, but the width of a trench (measured at the bottom) is not greater than 15 feet.

UNDERGROUND INSTALLATIONS: Include, but are not limited to, utilities (sewer, telephone, fuel, electric, water, and other product lines), tunnels, shafts, vaults, foundations, and other underground fixtures or equipment that may be encountered during excavation or trenching work.

UNCONFINED COMPRESSIVE STRENGTH: The load per unit area at which soil will fail in compression. This measure can be determined by laboratory testing, or it can be estimated in the field using a pocket penetrometer, by thumb penetration tests, or by other methods.

UPRIGHTS: The vertical members of a trench shoring system placed in contact with the earth and usually positioned so that individual members do not contact each other.

WALES: Horizontal members of a shoring system placed parallel to the excavation face whose sides bear against the vertical members of the shoring system or earth.

VIOLATIONS:

Serious: A violation in which there is substantial probability that death or serious physical harm could result, and the employer knew or should have known of the hazard.

Willful: A violation committed with an intentional disregard of, or plain indifference to, the requirements of the OSH Act and regulations.

Definitions that Are No Longer Applicable

For a variety of reasons, several terms commonly used in the past are no longer used in revised Subpart P. These include the following:

Angle of Repose Conflicting and inconsistent definitions have led to confusion as to the meaning of this phrase. This term has been replaced by **Maximum Allowable Slope.**

Bank, Sheet Pile, and Walls Previous definitions were unclear or were used inconsistently in the former standard.

Hard Compact Soil and **Unstable Soil** The new soil classification system in revised Subpart P uses different terms for these soil types.

Index